The origins of angiosperms are still debated, despite many years of work by scientists from differing disciplines. The progress made toward resolving the problem is reviewed in this book. The author suggests that the only fruitful method of study is the total integrated use of the fossil record, particularly dispersed palynomorphs. This includes the use of electron microscopy and refined data handling to record the occurrence of microscopic fossils, rather than the extensive use of morphology and cladistics. The methods advocated in this book could result in a rethink of the current classification of living plants. The ideas presented will initiate discussion between professionals and students of paleontology and plant science on the wider possibilities that may clarify the enigmatic origins of the dominant flowering plant groups.

Cambridge Paleobiology Series

1 The enigma of angiosperm origins

CAMBRIDGE PALEOBIOLOGY SERIES

Series Editors:

D. E. G. BRIGGS, University of Bristol
P. DODSON, University of Pennsylvania
B. J. MACFADDEN, University of Florida
J. J. SEPKOSKI, University of Chicago
R. A. SPICER, University of Oxford

Cambridge Paleobiology Series is a new collection of books in the multidisciplinary area of modern paleobiology. The series will provide accessible and readable reviews of the exciting and topical aspects of paleobiology. The books will be written to appeal to advanced students and to professional earth scientists, paleontologists and biologists who wish to learn more about the developments in the subject.

The enigma of angiosperm origins

NORMAN F. HUGHES
Department of Earth Sciences, University of Cambridge, and
Queens' College, Cambridge

CAMBRIDGE
UNIVERSITY PRESS

CAMBRIDGE UNIVERSITY PRESS
Cambridge, New York, Melbourne, Madrid, Cape Town, Singapore, São Paulo

Cambridge University Press
The Edinburgh Building, Cambridge CB2 2RU, UK

Published in the United States of America by Cambridge University Press, New York

www.cambridge.org
Information on this title: www.cambridge.org/9780521411455

First published 1994
This digitally printed first paperback version 2005

A catalogue record for this publication is available from the British Library

Library of Congress Cataloguing in Publication data

Hughes, Norman F. (Norman Francis)
  The enigma of angiosperm origins / by Norman F. Hughes.
    p.   cm. – (Cambridge Paleobiology Series 1)
  Includes bibliographical references and index.
  ISBN 0 521 41145 9
  1. Angiosperms, Fossil.   2. Paleobotany – Mesozoic.   I. Title.
II. Series.
QE980.H79   1994   93–41393 CIP
561'.2 – dc20

ISBN-13 978-0-521-41145-5 hardback
ISBN-10 0-521-41145-9 hardback

ISBN-13 978-0-521-67554-3 paperback
ISBN-10 0-521-67554-5 paperback

Contents

Preface

This book is intended as a sequel to my previous book published in 1976 (*Paleobiology of angiosperm origins*, CUP), but is not in any sense a second edition. The still unsolved problem of angiosperm origins has moved on in terms of interest and of general attention, but remains in some respects as wide open as ever. It is currently possible to publish serious papers proposing origins scattered over 150 million years of Triassic to Cretaceous time, a duration longer than the probable subsequent period of existence of this centrally important group of plants. Having studied Mesozoic fossils relevant to the problem over the last forty years, and having thus been personally involved along with many others in the failure to solve it, I sincerely believe that this failure is not due primarily to lack of application on the part of any one of us, and that consequently there is a significant error of method concealed somewhere.

The current divide appears to be between the relatively few who maintain that reliable evidence can only come from the fossil record, and perhaps the majority who feel that all the weight of scientific effort committed over many years to comparative morphology, to interpretation of chromosome numbers, to DNA or RNA signatures and the like should somehow prevail. To a believer in the ultimate efficacy, when properly employed, of the whole integrated fossil record, despite its shortcomings, the 'comparative morphology' method can offer only approximations with no time-sequence basis and thus no penetration beyond a fog of enthusiasm that is not meaningfully susceptible to statistical manipulation. Compromise has already been tried for perhaps two decades now without any noticeable success, and will doubtless persist, but the strong suspicion that the significant error of method is indeed to be found in this neontologic procedure suggests that compromise is not any longer of value here.

Unfortunately the issues very easily become emotionally charged. This is a measure of the great investment in the existing shored-up part-solution of the problem; the very central classification of extant angiosperms, leading on to good order in botanical science and even to public esteem, appears to hang on the maintenance of this uneasy position. Because the consequences of any change of approach in the direction proposed might reach into the understanding of evolution, into the classification of all land plants affected by concepts of primitive and advanced characters, into taxonomy and the nomenclature codes, and into recording methods in the whole of paleontology, there is great difficulty in avoiding the giving of unintended offence.

No unusual rectitude is claimed for the ideas presented here, and no rapid solutions are offered, but I hope that the challenge is serious enough to encourage exploration for new lines both of evidence and of argument.

Although I have trawled ideas both positive and negative from very many paleobotanical colleagues past and present throughout the world, most of whose names appear at some stage in the book, none of them should be held to account even by association for any disasters of mine. Particularly helpful have been some of the botanical brigade of free-thinkers of the unexpected such as Tom M. Harris, E. J. H. Corner, Jack Douglas, William G. Chaloner, Val A. Krassilov, A. D. J. Meeuse, James A. Doyle, John M. Anderson and others who have unknowingly quashed for me various wilder aberrations. Research students and associates who have suffered and thus helped, knowingly or otherwise, include the late R. Ashley Couper, Frank R. Gnauck, Kathleen I. M. Chesters, Margaret G. Mortimer, Geoffrey Playford, Mary E. Dettmann, Jenny K. Friend, Geoffrey Norris, Keith C. Allen, Judy C. Moody-Stuart, Elizabeth M. Truswell, David J. Batten, Catherine A. Croxton, David G. Smith, John F. Laing, Timothy C. B. Oldham, Gillian E. Drewry, the late Timothy H. Jefferson, Jenny L. Chapman, Babagide Salami, Hamish J. Campbell, Audrey G. McDougall, Ihsan I. Aslam, James H. J. Penny and Ian C. Harding. I have also been very grateful for long periods over the years for much technical assistance in the Department of Earth Sciences from the late Albert Barlow, Paul Hensher, Stan Curtis and David Newling, and from many others at particular times. In the last few months I have been specially appreciative of help from Robert A. Spicer, Alan G. Smith, Gillian E. Drewry, Catherine Flack and Sandi Irvine of CUP, Audrey McDougall, Christine Few, Jacqueline Hodkinson and Hilary Alberti.

Norman F. Hughes

Illustration credits

The author is grateful to the following publishers and journals for permission to reproduce illustrations. The full forms of the references are given at the end of the book.

American Journal of Botany: 8.56 (Taylor and Archangelsky 1985), 10.7 (Drinnan *et al.* 1991), 10.8 (Dilcher and Kovach 1986).

Cambridge University Press: 5.2 (Hughes 1989), 8.3A, 8.9E,F, 8.13B (Stewart 1983), 8.22, 8.24, 8.34, 8.41, 8.44, 8.46, 8.58 (Hughes 1976), 10.3 (Upchurch and Wolfe 1987), 10.5C (Friis *et al.* 1987).

Columbia University Press: 8.49A (Watson 1988).

Elsevier Science Publishers BV: 8.14A,C, 8.32, 8.49B-D (Alvin 1982), 8.19B,C (Pedersen *et al.* 1989), 10.4D,E (Krassilov *et al.* 1983), 10.9 (Krassilov and Pacltova 1989).

ETP Publications: 8.11 (Cornet 1989*b*).

Missouri Botanical Garden: 8.4B, 8.6B-F, 8.19A, 8.27, 8.29, 8.36 (Crane 1985), 10.5A,B (Dilcher and Crane 1984).

National Botanical Institute, Pretoria: 8.4A (Anderson and Anderson 1985), 8.6A (Anderson and Anderson 1989), 8.9A-D (Anderson and Anderson 1983).

Palaeontology: 8.20, 8.53 (Hughes 1961).

Scandinavian University Press: 10.4A-C (reprinted from Krassilov and Bugdaeva 1988*c*, by permission of Scandinavian University Press).

E. Schweitzerbart'sche Verlagsbuchhandlung: 8.3B-D (Pant and Nautiyal 1984), 8.14B, 8.15 (Jung 1968), 8.18 (Schweitzer 1977), 8.39 (Krassilov 1975), 10.6A-C (Retallack & Dilcher 1981*b*).

United States Department of the Interior, Geological Survey: 8.60 (Ash and Read 1976).

University of Chicago Press: 10.6D,E (Drinnan *et al.* 1990, © 1990 by University of Chicago. All rights reserved), 11.1 (Doyle 1973, © 1973 Stony Brook Foundation, Inc.).

PART I

The setting of the problem

1 Outline history of investigations

1.1 The general purpose of study

Angiosperms are the perceived group of living plants dominating most land areas of the present world in an astonishing variety of form and function. Their total number of species at more than 250 000 easily exceeds the numbers from all other groups of plants taken together. They naturally figure in most studies by botanists, and they form the basis of most pharmacology, horticulture and agriculture; their conservation and successful exploitation closely affected the fortunes and future of humanity, and of all other animals. In the geologic past, angiosperms clearly dominated all the lands in Cenozoic and Late Cretaceous time covering the last 100 million years, with presumably a similar influence over the land animals concerned. Geologists attempting to unravel the past conditions of land life on earth, employ studies of stratigraphy, paleogeography, paleoecology, and paleoclimatology, which are all greatly dependent in this era on plant fossil evidence provided from angiosperms. It is therefore logical for all the scientists involved in these studies to attempt to understand, to classify and to predict the properties of these universal living plants and the preceding fossils, by elucidating their evolutionary history; this investigation has, however, proved to be unexpectedly difficult.

1.2 The object of search

A definition of an angiosperm is an elusive target; the sum of characters of 250 000 living species cannot be otherwise and in practice there are no characters that are present in all of the extant species. The difficulty of definition could reflect a proposition that the living angiosperms are not a natural group but represent a level of development now attained by land plants with seeds. In practice, paleobotanists have stayed primarily with the characters that face them in fossils: (a) reticulate venation in the leaf (including the monocotyledon pattern), (b) columellate–tectate pollen, and (c) wood that includes vessels. A glimpse of an occasional more elaborate character, as in a flower, amounted to a bonus. Until late in the twentieth century supposedly critical characters for identifying angiosperms have often been seen rather negatively as very rare successes or as frequent failures to rise above gymnosperm level, the latter being defined more easily (than an angiosperm) from a relatively small number of living plants.

1.3   The nineteenth century

Plant fossils were widely and successfully studied through the nineteenth century, both in Europe and a little later in North America, where with few exceptions pure exploration continued until the end of the century. Relatively familiar-looking angiospermous leaves were common in Tertiary rocks, but in Europe these were separated stratigraphically from all Mesozoic and earlier rocks by the widespread White Chalk marine succession (barren of such fossils) so that no relevant evidence of fossil sequence was available for this critical Cretaceous period of the early development of angiospermous plants. At the time there was so much else to examine in paleobotany that little comment was made on this difficulty. In the first half of the nineteenth century perhaps more attention was paid to animal fossils that were often more complete and thus easier to interpret. Significant work on plants was largely concerned with the unusually well-preserved material from Carboniferous rocks and from adjacent Late Paleozoic strata. In mid-century, Charles Darwin (1859) remarked in a letter on the 'mystery' of the origins of angiosperms, but did not follow it up himself; others were probably more exercised by the principal elements of his work and did not at the time appear to have genuinely or directly accepted the challenge of this particular 'mystery'.

Very large numbers of observations of angiospermous fossils were made by Oswald Heer of Zurich (1809–1883) especially at Oeningen (1876) and elsewhere in Switzerland but also with collections from all over the then known geologic world. He apparently believed in a kind of creationist punctuated equilibrium and was not an admirer of the work of Darwin. Baron von Ettingshausen of Vienna also recorded angiospermous leaf fossils from all over Europe, and was an enthusiastic follower of Darwin, but his work was criticised as inadequately researched by Schenk (Professor of Palaeophytology at Leipzig 1868–1891), and subsequently by many others. The first complete textbook on fossil plants was attempted (1868–1874) by Schimper in Paris, and was eventually completed by Schenk, who exerted great influence on the subject and who insisted that a sound knowledge of systematic botany was essential. This dictum appears to have carried enough weight to have resulted in virtually all late nineteenth century paleobotanical practitioners being botanists. A consequence of this was that angiospermous fossils became neglected again because most botanists regarded leaf fossils as of no use to a botanical systematist. This is reflected in typical textbooks of fossil plants at this time. Solms-Laubach (1887; 1891 in English translation) gave only six individual index mentions and no chapter to angiosperms in a 400 page book. Zeiller (1900) devoted fifty pages to angiosperms, but very few leaves were illustrated and those were attributed to von

Ettingshausen; Zeiller did, however, pick up some monocotyledonous fossils and particularly palms from Late Cretaceous and Eocene strata. No attention was paid to possible origins of this vast group of angiosperms that dominate the Holocene floras.

This state of appreciation persisted into the first quarter of the twentieth century, when great advances were being made with Paleozoic plants and even to some extent with Jurassic plants by Nathorst and others. It presumably formed the background to the important attempts by Arber and Parkin (1907) and by Wettstein (1907) to visualise angiosperm ancestors on a purely theoretical basis. Seward's *Fossil plants* (1898–1919) in four volumes ceased after describing fossil coniferophytes; interestingly Seward did not attempt to cover fossil angiosperms in his other authoritative works, although a field trip in West Greenland (Seward 1926) did result in a geographic theory of origin of the angiosperms in the Arctic region. Many geologic observations since his time have tended through improved dating to invalidate this suggestion.

1.4   The early twentieth century

The transition to the twentieth century was gradual in paleobotany, with the science dominated by Zeiller (France), Kidston (UK), Nathorst (Sweden) and Seward (UK). In America, Wieland researched cycads and cycadeoids and Berry the description of the Potomac floras of Maryland, although the stratigraphic position of most American early Cretaceous rocks remained imprecise. In Russia, Zalessky and subsequently Krychtofovich were naturally concerned with exploration in that vast territory. The students of Seward, particularly Hamshaw Thomas (1925) using cuticles in the description of the new Caytoniales, and Sahni (1948), who ultimately described the *Pentoxylon* group from India, sought information on angiosperm ancestors.

These palaeobotanists were all botanically trained and one acceptable conclusion may be the repetition of Schenk's earlier dictum that a clear knowledge of, or at least an adequate feeling for, systematic botany is indispensable in this search, although now must be added the qualification that it should be accompanied by great restraint in applying such knowledge beyond the 'time' field in which it is based. A second conclusion or warning might well be that frustration and disappointment are likely in the absence of an equally sensitive appreciation of time-scale and stratigraphic succession in relation to the distribution of fossil discoveries.

1.5   Routine search and some despair

The twentieth century, however, was soon to witness a significant paleobotanical failure, which had the effect indirectly of discouraging attention to the angiosperm origin and various other problems. The failure was in data handling of the very large quantities of well-preserved and varied fossil plant materials of the Carboniferous Coal Measures, which were seen not to be providing the promised stratigraphic control that geologists expected of these valuable resources. With hindsight, the failure can be seen to have been due to very weak taxonomy unduly influenced by overbearing traditional nomenclature; the effect was creeping and was entirely international through the work of many diligent paleobotanists including Kidston (UK), Jongmans (Holland), Gothan (Germany), Bertrand (France), Crookall (UK), Dix (UK) and Remy (Germany). The disastrous result was that geologists working on the Carboniferous Coal Measures turned away for stratigraphic delimitation to the far less character-bearing non-marine lamellibranchs (bivalve molluscs), with which in the end they achieved very little more, while all fossil plants were completely left aside. This had happened in Britain by the time of 1939–1945 war, but the problem was earlier evident in the extraordinary wasted opportunity of David Davies (1929) in which great effort was expended on a huge collection of plants from one South Wales colliery without any profitable outcome, evidently because ideas were lacking. The whole of paleontology for geologists became almost completely centred on animal fossils for a generation.

Perhaps Hamshaw Thomas became discouraged after perceptively erecting both the Caytoniales and the Corystospermales but not being adequately acknowledged for his success with the problems of these fossils; again with hindsight, he was in effect defeated by the dogma of a monophyletic origin for the angiosperms. Subsequent prominent Mesozoic workers such as Florin (Sweden), T. M. Harris (UK) and Kräusel (Germany) all dealt with fossil gymno-sperms, but this was almost certainly to develop to the full the study of cuticles, which in this Mesozoic group were both thick and very durable and therefore newly useful as evidence of characters. Angiosperms were studied at this time but quite separately and essentially from a Cenozoic point of view by, among others, Chandler (UK) and Chaney (USA). Other well-known paleobotanists including Hoeg (Norway), Arnold, Andrews and Banks (all USA) and Leclerq (Belgium) were primarily concerned with the Devonian early plant life. Right through until about 1960, the only further angiosperm origin activity was a scatter of poorly substantiated claims based on individual characters of fossils and often on single specimens (for details, see Hughes 1976, chapter 13).

1.6   The advent of palynology

Although pre-Quaternary paleopalynology was born from German brown coal (Cenozoic) and then from Carboniferous studies and some early Soviet work in the thirties, most interest was generated by Schopf *et al.* (1944), in a simpli-fied and well-illustrated manual on palynomorphs of the North American Pennsylvanian Coal Measures. The fifties decade was a period of organisation led by Thiergart, Pflug and Robert Potonié (all Germany), Naumova (USSR), Dijkstra (Holland), Couper (New Zealand) and others; the classifications of palynomorphs, although necessary, became over elaborate and to some extent have since inhibited the general reconnaissance exploration of the sixties and later. Not until the mid-seventies, however, was paleopalynologic study applied directly to the angiosperm origin problem; general Phanerozoic (Phanerophytic) exploration had until then provided adequate general incentive for study.

1.7   *The current background*

Thus, until the last quarter of the twentieth century, there had been no significant new area of progress with fossils of possible angiosperm ancestors since the work of Hamshaw Thomas. Throughout the century, however, there had been great activity in the classification of living angiosperms including the work of Engler and Prantl (1897–1915), Wettstein (1901), Bessey (1915), Hutchinson (1964), Cronquist (1968), Takhtajan (1969 translation), Thorne (1976) and Cuerrier *et al.* (1992). Naturally because of the relative vacuum in studies of fossils, botanists interpreted phylogenies from consideration of these bountiful sources alone. Although such interpretations may have appeared promising for a while, it is now clear that they cannot replace the study of the fossil record in its relevant entirety and also clear that they have frequently misled scientists into work on diversionary topics.

2 Current contributions on origin

2.1 Trying all approaches

In the last fifteen years there has been more sustained interest in the angiosperm origin problem than ever before. Many paleobotanists and botanists have contributed either directly or in passing, and there have been many more suggested solutions, the majority based on observations of Cretaceous pre-Aptian fossils, at least in the first instance. The several recent general textbooks of paleobotany (Taylor 1981, Stewart 1983, Thomas and Spicer 1986, Meyen 1987, Taylor and Taylor 1993, Stewart and Rothwell 1993) comment on some of these suggestions but mostly follow tradition and devote surprisingly little space to either the development or the subsequent evolution of angiosperms. The exceptions are those of Thomas and Spicer (1986), who allotted nearly a third of their space to a balanced criticism and some useful suggestions, and Taylor and Taylor (1993), who updated all recent discoveries. A personal selection of major contributors to the solution of the problem is offered below, primarily to illustrate the diverse nature of the continuing difficulties; apologies are certainly due to any whose ideas have been unintentionally distorted, unduly abbreviated, or worse still omitted; obviously many more have thought about the problem.

2.2 Detailed development of comparative morphology

Among the more theoretical investigations was that of Kenneth R. Sporne (1915–1989), who very thoroughly compiled (Sporne 1980, also 1974 and earlier) comparative morphology characters for all extant angiosperm families into his 'Advancement Index' in which a low percentage score was considered to indicate primitive taxa. Monocotyledons had to be handled separately under different characters (see Sporne 1974), and a strangely assorted set including Liliaceae and Orchidaceae showed low (primitive) values, while palms, grasses and many water plants showed high values; no recognisable pattern emerged. Among dicotyledons, Magnoliaceae showed a low value of 25%, several Hamamelid families 33% (Sporne 1980), and also some Euphorbiaceae, but the pattern otherwise was of relatively high values and curiously the Chloranthaceae, so much favoured in various other schemes (see below), had an index of over 60%. Some of these values were varied by Chapman (1987a), who re-examined the statistical basis, and attempted some calculation of 'Evolutionary Distance'

(ED). Sporne himself did the original extensive compiling work with several updates but chose not to speculate; 'it is prudent to go no further than knowledge will allow' (Sporne 1974, p. 184), although his sympathy appeared to lie with Bessey (1915), with Takhtajan (1969) and with Cronquist (1968).

2.3   Application of various evolutionary theories

Sergei V. Meyen (1936–1987), well known for his encyclopaedic work on Permian and other earlier fossil plants, has suggested that a bennettitalean ancestry for angiosperms could appear more botanically probable if allowance were made for gamoheterotropy, the transfer of characters in evolution from one sex to another (Meyen 1988). In detail he proposed the origin of the angiosperm gynoecium from the bennettitalean microporophyll structure as support for similarities already observed in wood, stomatal structure and pollen. Sadly, this paper was posthumously published and Meyen was not able to develop the idea further, particularly in relation to the foliage of the two groups.

William C. Burger provided a brief study of the success of angiosperms in their very great variety (Burger 1981a), followed by a lively revival (Burger 1981b) of an old hypothesis that monocotyledons arose first; although he does not discuss reproductive structures as such, this revival includes the slight possibility of monocotyledon origin even from eusporangiate pteridophytes. He refers in passing to Triassic *Sanmiguelia*, but this was prior to the publication by Cornet (1986). The introduction of his paper (Burger 1981b, pp. 189–90) eloquently presses the case for an alternative theoretical framework for the origin of angiosperms to avoid circularity of argument inherent in matching data to a single overriding hypothesis such as a monophyletic origin from gymnosperms through dicotyledons to monocotyledons.

A. D. J. Meeuse announced himself as a dissident (Meeuse 1990 and earlier) and his non-acceptance of what he referred to as traditional foliar carpel theory led to his Anthocorm Theory (Meeuse 1981, p. 434) of a central axis becoming a group of bracteated gonoclads. Unfortunately for the author he had developed the theory through various improvements and amendments over twenty years in a long succession of earlier papers, which he subsequently considered with some truth to have confused his readers. Additionally, although he did not himself work with fossils, Meeuse (1979) assessed with very great enthusiasm all the Mesozoic literature and fairly concluded that fixed attitudes in particular to a monophyletic origin of all angiosperms were responsible for the marked lack of progress and to human-imposed barriers of distinction between 'gymnosperms' and 'angiosperms'. Although many botanists may not appreciate his ideas, his stirring contribution has provided very important diversity of approach.

David L. Dilcher, although his better-known contributions are to new Cenomanian plants from the Dakota Formation, provided (Dilcher 1979) an encouraging and stimulating discussion of some of the angiosperm origin problems.

While firmly highlighting the fossil record, he described the long established 'primitive angiosperm' concept as biased and thus unhelpful. On the interpretative side he suggested that the small flowers in such groups as the Hamamelidales should be considered as initially simple rather than as 'reduced' flowers, again supporting a straightforward uncomplicated approach.

2.4   Relatively remote pre-Jurassic fossil evidence

Working with Early Carboniferous pteridosperms, Albert G. Long (1977) followed botanical tradition in regarding the angiosperm origin problem as essentially concerned with the carpel, enclosed ovules, and the evident difference of inner from outer integument. He was comparing the excellently preserved Lower Carboniferous *Eurystoma*, which he had found and developed, with an idealised 'central' angiosperm. This evidence should certainly be added to that from other very early pre-Cretaceous records, but it carries on its own no special conviction.

Working in the fifties in the Herbarium of the Royal Botanic Gardens, Kew, on floral evolution, Ronald Melville (1903–1985) developed his theory of angiosperm origin through the 'gonophyll' rather than through some kind of carpel, which he rightly insisted had not been demonstrated in any pre-Cretaceous fossil. Melville (1962, 1983) sensed the importance of *Glossopteris*; he was particularly interested in the then-supposed bisexual nature of the fructifications of these fossils which were being described essentially for the first time (Plumstead 1952). His enthusiasm appeared to lead him to describe the anastomosing leaf venation pattern of *Glossopteris* as discernible in many angiosperm petal and leaf details, and to depend on descriptions of *Glossopteris* fructifications made for the most part from very difficult impression fossils without cuticle or even carbon (see also Retallack and Dilcher 1981a). Once again it must be said that these suggestions should certainly be listed, but that this very early evidence alone is not convincing. It should also be recorded that Melville (1983), in defending his proposal, was obliged to admit that all *Glossopteris* fructifications then known were unisexual, and that is still the position (Taylor and Taylor 1993). Additionally in explaining the time gap of appropriate fossils by invoking Earth Magnetic Reversal to account for the 'sudden' destruction of the *Glossopteris* flora and replacement by the *Dicroidium* flora, and by invoking upland survival (unobserved) through the Jurassic, his hypotheses became more and more elaborate and unlikely. The original morphologic point, however, should stand in the record as a possibility.

After greatly expanding the knowledge of *Sanmiguelia* (Cornet 1986), a unique Late Triassic plant first discovered in Arizona thirty years earlier, Cornet (1989b) has described a large new palynomorph assemblage from the Late Triassic beds of the Richmond Rift Basin, Virginia. In both cases he has enthusiastically introduced evidence for Triassic angiosperms. The megafossil reproductive organs of *Sanmiguelia* are unlike those of any other described plant, angiosperm or gymnosperm. From Virginia 2% of the dispersed pollen, among predominantly normal late Triassic forms, includes eleven new species of angiosperm-like reticulate-columellates in a new group of 'Crinopolles'; these new species broadly resemble Cretaceous *Liliacidites* but they are of considerably larger dimensions. Although many workers will await some connecting fossil records from Jurassic rocks (see Cornet and Habib 1992), all this new evidence clearly requires assimilation into those theories that at present include first effective radiation in early Cretaceous time.

2.5   Cryptic upland vegetation

Writing from south-western Australia, which has a startlingly rich extant flora of angiosperm endemics, J. S. Beard (1989) challenged Retallack (1977), who was also followed by White (1986), over his belief in extensive Mesozoic upland vegetation; Beard found no evidence of plants living far from water, evidence that might have been provided by adequate descriptions of the nature of fossil soils of the period, if they had been found. He further pointed out that rafted logs of trees are not evidence for upland forests; they are most likely to represent river-bank vegetation ripped away by floods, as upland tree casualties do not on average travel far. For other reasons he also suggested that it was the angiosperm radiation itself that first began to clothe the non-aggradational land, and that perhaps not immediately in Cretaceous time.

As author of an influential overview of the whole group of flowering plants Armen Takhtajan (1969 and earlier *in litt.*) appreciated the lack of progress either through comparative morphology or fossils and so attempted to apply the theory of neoteny (Takhtajan 1976 and earlier) borrowed from a previous generation of zoologists and animal geneticists. He envisaged this throughout the vascular plants and particularly in the derivation of Magnoliophyta from gymnosperms, and Liliopsida (monocotyledons) from Magnoliopsida (dicotyledons). He described the effect in detail, particularly in the case of the apparently reduced form of both male and female gametophytes. The logic continued to the point of concluding that 'neotenous forms arise under some kind of environmental stress', and that in the case of early flowering plants it probably consisted of 'moderate drought on rocky mountain slopes in an area under monsoon climate'. This amounts again to the cryptic evolution in upland areas, an overelaborate proposition that most geologists find little reason to support.

2.6   General study of gymnospermous fossils

Tom M. Harris (1903–1983) was a great investigator of Mesozoic plant fossils both in the field and in the laboratory, (e.g. Harris 1961, 1964, 1969, 1979;

Harris *et al.* 1974) and he eschewed speculation in his penetrating descriptions of many 'unexpected kinds of plants, each a problem in itself' (Chaloner 1985). He was directly involved (Thomas and Harris 1960) in the elucidation of *Caytonia*, and therefore automatically drawn into the angiosperm origin problem (Andrews 1980). Despite his curious reluctance to countenance any new taxa for fossils (Harris 1976), his records of Mesozoic seed-plants remain the best available in very many groups and consequently most relevant hypotheses will come to be tested against his knowledge and expertise.

2.7   Earliest angiospermous megafossils

The diverse nature of Mesozoic seed-plants has frequently been emphasised by descriptions of new discoveries from Asia published in Russian and frequently also very lucidly in English by Valentin A. Krassilov (1977, 1982a, b, 1984, 1987, 1990, 1991). He maintains a healthy scepticism about most dogma, both botanical and geologic, and particularly supports a polyphyletic origin (Krassilov 1990). Like others from the USSR he preferred the era-scale geologic time division Mesophyticum from approximately mid-Permian to mid-Cretaceous periods (see Chapter 14) as being more naturally relevant to events on land, but this useful Russian device has not been followed elsewhere; this may be because the general trend to greater precision causes geologic periods rather than eras to be more often used as time divisions except in very sweeping reconnaissance. Relatively new observations on interaction between plants and insects are recorded from early Cretaceous rocks of Mongolia (Krassilov and Sukatcheva 1979) and Transbaikalia (Krassilov and Rasnitzyn 1982), and comments on vegetation changes and extinction of dinosaurs (Krassilov 1981). All these papers are attractive for a wealth of new ideas and asides, sometimes heretical and frequently not developed to the full. He has also brought together in a book a great deal about the paleobiological events of the Cretaceous period (Krassilov 1985).

Subsequent to his collaboration with Doyle, Leo J. Hickey has published articles on the major features of angiospermous leaf architecture in the fossil record (Hickey 1978, Hickey in Metcalfe and Chalk 1979) and particularly of their origin in mid-Cretaceous floras. Others have subsequently built further on this information in the continuing struggle to relate Cretaceous vegetation and climate (Upchurch and Wolfe 1987). Also relating to probably mid-Cretaceous paleoecological conditions for early angiosperms are two papers (Hickey and Doyle 1977, Hickey 1984), of more geologic discussion.

2.8   Earliest obviously angiospermous pollen

In addition to being the first (Brenner 1963) to describe fully an important pollen-bearing succession of Mid-Cretaceous age in Maryland, Gilbert J. Brenner (1976) was also the first to relate occurrences in different Cretaceous provinces

from well co-ordinated palynologic data. He was further concerned in the discovery of Hauterivian angiospermous pollen from Israel (Brenner 1984), and
even earlier records also from Israel (Brenner and Bickoff 1992).

James A. Doyle was originally concerned with the dispersed pollen of the
Potomac Group in the eastern USA, where an unconformity cuts off the earliest
part of the Barremian record (Doyle and Hickey 1976). However, he has also
published with French colleagues on Equatorial Africa (Doyle *et al*. 1977) and
'Northern Gondwana' (Doyle *et al*. 1982), where the succession goes much
lower, although precise stratigraphic correlation is less certain. In these and
numerous other papers there were figured and discussed many new pollen
types, and a theory of origination and early radiation from paleotropical regions
(Doyle 1978*a*). For twenty years his work has been central in early Cretaceous
plant studies and he has also neatly summarised his views (Doyle 1984). More
recently he has published on cladistic analysis of all major seed-plant groups
(Doyle and Donoghue 1987) to establish the most probable angiosperm
ancestors as close to Bennettitales, *Pentoxylon* and Gnetales; he has throughout
shown strong support for a monophyletic origin of angiosperms.

Renowned for their fine preparation technique and excellent scanning electron microscopy (SEM) and transmission electron microscopy (TEM) photographs of even isolated grains of dispersed fossil pollen (Walker and Walker
1984), James W. Walker and Audrey G. Walker also predicted in the same paper
(1984) and earlier the finding of pre-Barremian 'Lower Magnoliidae' despite the
probable difficulty of identifying such fossil pollen. They suggested that the
well-known Barremian monosulcate columellate–tectate pollen from the Potomac Formation and from England represented the second phase of development
of the plants concerned.

Following an earlier review of the whole topic (Hughes 1976), Norman F.
Hughes has exlored a relatively complete and well exposed available succession
in southern England of appropriate rocks from Late Jurassic to Albian age
(Hughes *et al*. 1979, Hughes and McDougall 1987); the Hauterivian–Barremian–
Aptian occurrences of essentially monosulcate, columellate–tectate (MCT)
pollen, are elaborated in Chapter 9.

Although others have not recorded such pollen in the Berriasian and Valanginian rocks, which have been studied for some years in England, Germany and
elsewhere, Lavinia Trevisan (1988) has found pollen of this type, together with
dateable dinoflagellates of Valanginian age from Tuscany in western Italy. No
megafossils, or indeed mesofossils, have yet been discovered to match this dispersed pollen or to indicate its provenance, but it is recognisably of MCT types
and extends their range.

2.9   *General fossil pollen synopsis*

From long tropical service in both botany and geology, Jan Muller (1922–1983)
provided a uniquely well-informed review (Muller 1981 and also earlier *in litt*.)

of Cretaceous and Tertiary fossil pollen occurrences that had been claimed by others to represent extant taxa of plants. He categorised all known records as 'adequate', as not proven, or as 'pending' (of which there were many), and a few as 'rejected' because of errors; his summary showed only one or two orders and no clear families of Aptian to Cenomanian age, a small but increasing number from Turonian to Campanian time, a climax of diversification in Maastrichtian time (across the K–T boundary, although he did not refer to this) and steady introductions thereafter through the Cenozoic.

2.10   Concluding note

Deliberately excluded from this chapter, but not of course from consideration elsewhere, are description and speculation based on post-Aptian fossil records, which include generally accepted angiospermid characters. The great variety of approaches mentioned above all explicitly raise points about earlier fossils or involved earlier material in the suggestions presented.

3 Challenges to customary procedures

3.1 Introduction

The evolutionary origin of angiosperms has been under intermittent investigation for so long without any striking success that it seems necessary to examine first some of the routine assumptions made by biologists and by geologists working on this topic. The points selected below as headings are by no means all independent of each other but the order has been chosen for ease of presentation and to reduce repetition. They may be divided into: a group of five points (2–6) concerning the handling of extant plants but amounting to background pattern; four points (7–10) that concern the past history of plant groups but are biological in content; and four more geologically based matters (11–14) that have sometimes misled authors because of interpretations that remain disputed in some degree. None of these challenges amounts to a great issue in its own right; they may all be described as customary attitudes and it is perhaps best to consider them piecemeal, with the possibility of freeing at least one lead at a time.

3.2–3.6 Handling of data from extant plants and fossils

3.2 Use of characters

Although selection of a part of a plant as an organ bearing a stated character is an undoubted convenience in description and even in some interpretations, the process relates primarily to the purpose of the observer rather than to the plant itself. 'Character' is a property (of a species) continuous throughout the organism concerned, and subdivision of such a 'character' or weighting of parts of it can only be artificial, even if considered in terms of a genetic or molecular explanation. Consequently the selection of a flower or of any other comparable single part structure as definitively representative of a living plant group should not be countenanced as meaningful in stratigraphically observed evolution through geologic time; this does not question the probable validity and certain convenience of such selection in construction of a routine classification of strictly contemporaneous objects at any chosen point in that time. In practice the use of the detail of the closed carpel or of any associated selected set of

characters defines a Holocene angiosperm but has no meaning in Early Cretaceous time because there are normally no supporting observations.

3.3   Comparative morphology

While there remained a serious lack of evidence for details of ancestor plants
in past geologic periods, it was understandable that neobotanists surrounded
by very large numbers of living plants should try experimentally to arrange
these plants in series of states from those apparently simple to others of much
greater elaboration. It was a short step to relabelling these states as respectively
primitive and advanced, and another short step apparently to believing that the
series in some way represented the course of evolution.

Sporne (1980 and earlier) used the maximum number of characters available
to him and successfully produced an 'Advancement Index' for all extant families
of dicotyledons, but he was always careful to avoid the next step of speculation.
Others have gone further (Taylor 1991, Taylor and Hickey 1992) and some
have suggested that molecular studies will enhance comparative morphology
to such an extent that these speculations about evolution will become progressively more accurate; Martin *et al.* (1989) claimed angiosperm origin at 300 Ma
solely on such comparative DNA evidence. It seems unlikely, however, that
molecular studies of this kind will achieve anything more than a great extension
in comparative morphology per se beyond the work covered by Sporne and
others. Such studies, despite their apparent sophistication, are almost certain to
remain wholly speculative until they are firmly supported by stratigraphically
arranged fossil evidence. The recent discovery (Golenberg *et al.* 1990) of surviving DNA in a Miocene magnoliaceous leaf fossil, and possible experiments
on it, will extend knowledge and raise hopes but will not change the overall
restricted nature of the evidence.

3.4   The genus

The genus, ever since Linnaeus' time, has been usefully coined as an exploration
device (the binominal) to bring some order into great numbers of records of
random discoveries. It has more recently, in the era of nomenclatural codes
such as the International Code of Botanical Nomenclature (ICBN), become a
nomenclatural device and thus unavoidable in indexing. At the same time the
genus has quietly been credited with also having a biological meaning of its
own, in a way that has even spread from biology into paleontology; it has been
impossible to pin this down into any kind of quotable definition, although this
lack apparently does not diminish belief in such a meaning. Unfortunately and
quite without intention the genus as a nomenclatural priority device and the
rules surrounding it have become a serious cause of instability of many com-

monly used living plant names (Hawksworth 1991). The effect on names of fossils is similar but a less urgent and severe matter because horticulture and commerce are not involved.

The genus can be said in only a very few special cases to represent a past gene-pool, and does not normally amount in any sense to a present one among living plants. No new species in evolution can be shown to be descended from a genus, but only from an individual species (with a generic epithet to its name). The number of species in a genus can differ by two orders of magnitude; in the cases of the higher numbers the genus does not thereby produce any meaning in biology, and certainly no convenience at all in the handling of data. For such a unit, it can be argued that provision of convenience for the observer or for the systematist in being able to encompass all the included species for a study should result in a sensible general number limitation on species covered by such a nomenclatural device; this has not happened because of the persistence of belief in a meaning.

3.5   Higher taxa

In recent years higher taxa than genera have been manipulated with ever-increasing enthusiasm and zest, especially by purveyors of cladistic methods. This is intended to achieve more quickly, with more manageable data matrices, the results that have not flowed easily from past effort put into general comparative morphology. Among many others, Doyle and Donoghue (1987) have even woven higher groups of fossil plants into such manipulations but without apparently using the full data of their distribution in geologic time. Families and other ranks of higher taxa are even more remote than genera from biological meaning, other than in classificatory satisfaction, and they are usually nested with hierarchies that proliferate automatically to simulate higher organisation and to fill apparent gaps in the pattern. These absurdities are most obvious in cases of single isolated living species such as that of *Ginkgo biloba*.

3.6   Parsimony in nomenclature

Since early in the history of paleontologic exploration, it has been considered quietly meritorious to indulge in 'lumping' of observation records under one name, and correspondingly weak (indecisive) to defer to a common human failing in 'splitting' or promotion of many new names in face of uncertainty. T. M. Harris, probably the most effective worker on fossil plants in the last few decades, prided himself (Harris 1976) as a botanist in holding the line in this sector. Despite recent great advances in automation of data handling to relieve the human memory, this perversity of 'lumping' still persists in many individual attitudes. Even the general philosophical point that the detail of a phenomenon 'lumped' is lost to subsequent retrieval has not lessened the irrational fear of encountering too many names. Equilibrium could be reached by encouraging 'splitting' in the field of recording basic occurrence data, balanced by 'lumping' in any ephemeral interpretation that did not thereby affect the basic material. Attempting to organise both requirements for fossils within one taxonomy and one code of botanical nomenclature leaves no scope for a scheme to meet both needs separately as they arise.

3.7–3.10 Organic evolution and fossils

3.7   Phylogeny

Haeckel (1866) defined phylogeny as a 'branch of morphology which traces descent by means of homologous characters'. This is satisfactory and begs no questions, especially in zoological paleontology, to which it was first applied. Since then, what has apparently happened among evolutionary biologists concerned with botany is that 'morphology' has become 'comparative morphology' which has then become 'morphology of extant plants' because so very few fossils are adequately known. From this, many evolutionary biologists use phylogeny to mean the history of the development of a race, but they have permitted inclusion of many cases in which no historic evidence (of fossils) has been used or has even been sought. 'History' is therefore being used as a hypothesis only, lacking any possibility of proof and not seeking any. This bizarre situation is behind much printed work on evolution and the self-deception of both author and reader is deep enough to evoke even a slight sympathy with some of the objections of creationists concerning 'evolution', although none with their general creed.

3.8   The drive to 'identify' fossils

A fossil identified with something known before is a comforting step toward instant use of the new fossil to interpret paleoenvironment or stratigraphic correlation. Unfortunately haste to reach such conclusions often leads to dismissal of small doubts and discrepancies, which are then lost in an identification that it is thought cannot be both qualified and still remain 'useful'. Once again it appears desirable to separate into two distinct operations (a) the recording of data into which devices to express clearly the degree and direction of doubt can be incorporated, and (b) interpretation in which doubt needs either to be suppressed or to be more constructively expressed by dismissing the interpretation.

Again at an entirely different level all Mesozoic fossil plant genera appear to have been fitted if possible into known groups. A major extinct well-described group such as Caytoniales or Bennettitales is clearly acceptable, but if the leaf resembles a *Ginkgo* leaf or a *Taxodium* shoot there appears to be a very strong urge to ignore the lack of all other organs of the virtually unknown plant concerned and place it at once in a family Ginkgoaceae or Taxodiaceae. This process operating over a century or more has resulted in virtually all non-exotic Mesozoic plant fossils (the majority) being classed as cycads or cycadophytes, conifers or ginkgos. This has happened despite expert witnesses such as T. M. Harris constantly underlining the strangeness and surprisingly detailed properties of most of the fossils he encountered. At this higher group level the effect has been to narrow thought and imagination, even toward suggesting a certain drab uniformity in Mesozoic floras. Needless to say this view is hypothetical and has itself deadened initiative in seeking angiosperm ancestors. An ever-present example throughout virtually the whole of the Mesozoic is that of the plant or plants producing *Classopollis* pollen. Some Jurassic species of the conifer-like leaf *Brachyphyllum* produce cones with the pollen *Araucariacites* that are acceptable among 'conifers', but others produce the pollen *Classopollis*, which has no parallel in the 'coniferophytes' (see also Chapter 8). Identification is so beset with hazards that it ought to be a far more cautious, less popular and less automatic procedure.

3.9   Monophyletic models

The most satisfactory way to demonstrate monophyly is surely to assume the opposite as a general case and then steadily prove the single origin of the selected taxon through fossil evidence. Sadly the plain assumption of a monophyletic origin, applied particularly to angiosperms but also to other important groups, is almost automatic among the majority of botanists and paleobotanists. The principal traditional basis for this choice for angiosperms is the eight-cell female gametophyte, which is said to be uniform throughout the whole of the very large angiosperm group, and is thought unlikely to have originated more than once in this same form. An alternative explanation might be that such an arrangement of cells could be in necessary conformity with closed-carpel angiospermy itself and therefore to be expected in all such plants in some fairly closely similar form. Such a fundamental problem calls for a completely open choice of hypothesis in a matter that is likely to decide the whole manner in which ancestors are sought and checked.

3.10   Natural survival

This expression describes the probable process better than the 'natural selection' of hallowed origin, certainly for all plants and probably for most animals.

Reproduction in most instances provides very large numbers of progeny or propagules, of which nearly all individuals are destroyed or lost by natural causes. Survival of one or two individuals can be credited to chance by physical accident or to possession of the precise 'character' attuned to such persistence. To label this process 'selection' ascribes choice to what are in general mere physical forces or activities; this is little better than the common and misleading use of the word 'adaptation', which apparently credits choice or direction to the organisms themselves and which must be out of the question for most animals, let alone plants. So great are the losses inflicted on the very high productivity of most plants that the presumed interesting results of genetic mutation and diversity are very largely wasted. The argument has perhaps been overinfluenced by the experience of humans and other higher mammals in which survival rates are arranged to be much higher, but has virtually no place in dealing with naturally occurring plants.

3.11–3.14 Geologic and paleontologic understanding

3.11   Earth evolution

Organic evolution is frequently presented as a phenomenon on its own, about which for historical reasons extreme views may be engendered. It is more correctly to be regarded as a minor subset of all-embracing (physical) earth or even cosmic evolution; the geologic impermanence of every part of the crust, and of the mantle and core of the earth, is apparently little appreciated by most humans, nor is the position of the earth on a longer time-scale in the cosmos. It may well be the fault of geologists in failing to present this state of constant change clearly that has allowed biologists to claim excessive importance and mystery for their own small part of the whole.

The nature of geologic sedimentation and its products has changed with changes of water distribution and of development of salinity of the sea, with progressive changes of chemical composition of the earth's atmosphere, and with rearrangement of availability of less common chemical elements in the crust. Organisms, and particularly plants, have played their part in modifying directly the distribution of coals and petroleum and methane. Indirectly they have been involved in the evolution of soils through the slow development of true water-seeking roots from basal organs of plants, many of which in the Devonian period had no such function because the plants of that period probably did not extend beyond immediately available water.

3.12   Animal and plant integration

Only in the last two to three decades has interest developed in examining the question of plant and animal interdependence in past periods on land, and similarly even more recently in the sea. Direct evidence is rare and much circumstantial detail has to be used. So few animal groups have invaded the land (vertebrates, arthropods, gastropod molluscs, some 'worms' and some protozoa) that in Paleozoic time interaction must have been minimal; vertebrates were still carnivorous, and the invertebrates at that time used mainly dead plant material (trash) for food. From this, by Cenozoic time, cases of very elaborate plant–animal interdependence were set up, leaving the Mesozoic era as one of significant and far-reaching but gradual evolution of such relationships. Even the difficult interpretation of the gradual evolution of soils is not helped by the use for Mesozoic time of a Holocene classification of soils that is itself dependent on Cenozoic plant history.

3.13   Cryptic upland evolution of organisms

To explain long gaps in apparent evolutionary successions between ancestors in perhaps the Carboniferous or the Triassic period and sudden appearances in Cretaceous time, some biologists have postulated interim developments of life in upland regions from which almost no fossil record could be expected. Stratigraphers and other geologists have been insufficiently severe on such untestable hypotheses, supposing probably that the flaws were perhaps biological. Migration of some plants upwards to higher altitudes is an acceptable part of radiation from crowded and biologically competitive lowland areas on standard aggradational land on the lower courses of rivers; any return of upland survival plants into the thick of lowland competition is most unlikely on any large scale. Possibly geologists have misled others with discussion of relatively dramatic rises of global sea-level in the anti-glacial periods such as the mid-Mesozoic; over geologic time in millions of years sea-level rises were obviously gradual and slow, with continuous consequent changes to the distribution of aggradational land that sometimes, and perhaps usually, adjusted as rapidly as sea-level changed. Lowland habitats shifted geographically but were seldom if ever completely destroyed, being dependent on the continued flow of the rivers carrying sediment.

Climate changes, particularly in Mesozoic time, were mainly in the degree of latitudinal contrast. Holocene times, being glacial, provide extreme differences from tropics to poles; Jurassic and Cretaceous climates probably displayed much less contrast and could not therefore be as freely invoked as strong causes of plant migration.

3.14   The fossil record

Fashion and disappointments have engendered a long succession of complaints about the weakness and inadequacy of the fossil record, but it is important to recognise the word 'inadequacy' as being entirely anthropocentric. The fossil record is now well enough understood to allow prediction of the nature of the future successes that will continue as overall appreciation and new techniques develop. Even in plants, although fossil palynomorphs have already begun to be more fully exploited, there is a large relatively untapped field of mesofossils yielding small elements and fragments readily interpreted with the aid of electron microscopes. Better understanding of sedimentation leads to better targeted search for such materials. The fossil record, assuming continuing perceptive and diligent study, is flourishing and only needs to be accepted for what it is without complaint (Paul 1985). Hypotheses that can be tested only by impossible or most unlikely kinds of fossil discovery are best ignored. Fortunately, as ever, very occasional remarkable preservations still appear after diligent search, with resulting encouragement to all concerned; these special preservations may confirm suspicions or raise many more new problems but they can seldom be repeated. They represent local bonuses but the effect is statistically small on general interpretation, which has to depend on the general widespread occurrences of microfossils, and on data-handling methods still relatively undeveloped.

3.15   Summary

It can be said that all of the issues raised above contribute to the unrewarded present state of the search for angiosperm ancestors; the conventions to which they lead surround the study of Mesozoic land organisms in such a rigid manner that innovations appear unlikely. Some change in the surrounding structure is needed to provide freedom, even of thought. The challenges indicate a lack of commonly occupied ground between phytobiology and geology; for too long each side has politely assumed that greater wisdom in contentious matters lay beyond the divide in the plain where the greater experience was presumed to lie. The angiosperm origin problem has remained a victim of this empty area and clearly cannot be solved without genuine co-operation. To achieve this it is necessary to challenge modes of thought that arose round other different problems but remained as unplanned obstructions encircling this one.

4   Working principles

4.1   Introduction

This is an attempt to examine afresh the whole problem of the origins and early evolution of angiosperms, and to think around the many approaches and methods hitherto employed without any great or visible success. The following list of ten principles in use in this book is intended to assist in clarifying the choice and arrangement of topics in the subsequent chapters, and to act as a checklist if and when the argument may appear to have wavered. Reference back will be made in this context.

> TEN WORKING PRINCIPLES
>
> 1. Angiosperm origins comprise a problem, not a mystery.
> 2. Acceptance of the primacy of the fossil record.
> 3. Equal importance of all records of fossils.
> 4. Use of the unemendable paleotaxon for fossils.
> 5. Necessity of stratigraphic succession in all morphologic series.
> 6. Application of the simplest model of organic evolution.
> 7. All Mesozoic seed-plants are 'Pollenifera'.
> 8. Angiosperms are polyphyletic until proved otherwise.
> 9. Classifications of fossil plants are for single periods only.
> 10. Unique nature of the Cretaceous anti-glacial climatic maximum.

The reasons for selection and for application of each principle are discussed in the same order in the following paragraphs.

4.2   Problem not a mystery

The Mesozoic origin of angiosperms is a straightforward problem to be solved by diligent study of the fossil data known to be available. It is my contention that this problem has not yet been attacked in a sufficiently neutral manner clear of all identifiable prejudices; many of these have been unsuspected and undetected failures of method by biologists, by paleontologists (of which I have been one) and by geologists. This has arisen mainly because of numerous understandable attempts to find short-cuts around what appeared to some to be a long, unrewarding and indefinite task. The attitudes leading to such haste can be attributed to four simple human failings:

(a)   Impatience with fossil evidence, which takes special skills and much time to collect and to digest, and the consequent reliance as a substitute on a use of extant comparative morphology, which in this context amounts to guessing (see Mabberley 1984).

(b)   Reluctance to vary patterns of work over classification systems that have appeared satisfactory for other purposes, and have been barely questioned again in this context; the result is a rigidity of framework that replaces thought and predetermines narrow answers.

(c)   Inaccuracy in continuing to use unjustified and confusing anthropocentric vocabulary such as 'adaptation', 'phylogeny' and 'filling of niches', when such expressions can have no meaning in a context of pure natural survival.

(d)   Insensitivity or arrogance in regarding the study of evolution as a purely biological matter, and in the overconfident use of statistics on inappropriate databases, as in recently applied cladistics (Patterson 1981, Donoghue and Doyle 1989, 1991).

In some other sectors of paleontology, the results of comparable evolutionary studies have yielded some acceptable conclusions usually based to a considerable extent on fossils, but with angiosperm origin the combined effect of the many prejudices has led to the failure being unnecessarily regarded as a continuing mystery with an '*ex cathedra*' notation.

4.3   Acceptance of the fossil record

The fossil record, whatever its apparent deficiencies, is the *only* valid evidence of past events. In the case of land plants of the Mesozoic era, the record consists primarily of an 'ocean' of dispersed palynomorphs that has so far been very lightly assessed or even sampled; this ocean is studded with a very few 'small islands' of megafossils that are important but are inevitably unrepresentative and restricted in scope, and whose numbers are most unlikely to multiply significantly. Consequently the dispersed palynomorphs will form the main record, which must be developed and massaged to the full; much study already published is at a primitive reconnaissance level with very little SEM examination or quantification. Integration of data from different sources is in its infancy.

4.4   Equal importance of all records

It is necessary for all observations of fossil specimens to be retrievable in their entirety from the database. Unfortunately, in the past, very many observational records have been obscured in undescribed and unquantified attribution to previously described taxa, with the effect that nothing can be retrieved beyond an author's confident assertion of identity expressed in a species list. The construction of taxa should be regarded as a secondary activity to the making of a re-usable record; consequently the style of taxonomy forced on an author (Jansonius 1992) by adherence to the nomenclature requirements of the ICBN is no longer acceptable automatically in compiling a database. As an alternative, graded comparison records (see Hughes 1989, chapter 4) are necessary as a method of retaining for retrieval all the original observed information.

4.5   *Use of the paleotaxon*

Base taxa of fossils (known as paleotaxa), of level equivalent to that of the traditional species, are required as immutable descriptive points of reference in construction of an adequately accurate database and language of communication. Emendations of species, which are inevitable in the progress of science more than in the correction of error, have long been a source of confusion of reference; although accurate logging of emendations under the ICBN is allowed for and is possible, it is unhelpfully cumbersome and is seldom attempted in full because the system lacks any clear notation for that purpose. Paleotaxa differ from species in that they cannot be emended in any way (see Hughes 1989, chapter 6), and are consequently replaced by newly described further paleotaxa when new information becomes available; they are uniquely identified by author and date, in such a way that no priority system for names nor any centralised control is necessary. The name used is a binominal for compatibility with the traditional system, although in place of a generic epithet the title of a division of the Global Stratigraphic Time-Scale (known for this purpose as a 'timeslot') is used; a neutral generic epithet could equally well be employed if preferred. Higher taxa when necessary at all are kept at an informal and minimal level to avoid the development of unnecessary and meaningless hierarchies. For past records, names of species (and genera) are maintained in parallel with any new paleotaxa, as the form is compatible. Graded comparison records may be used with either form.

4.6   *Stratigraphic succession*

Only those sequences of observation records that can be shown by correlation to be placed in an ascending stratigraphic order can be admitted as evidence of any evolutionary succession of past organisms. Although this requirement may appear absolutely obvious to a geologist, the number of past papers from botanical sources that discuss morphologic series or proposed morphologic series without regard to this principle is too great to consider quoting; the stratigraphic occurrence level of important megafossil finds is seldom made clear, even in paleobotanical textbooks, and its absence seriously affects many optimistic references to angiosperm origins. The constraint when applied also usefully eliminates the mention of all later stratigraphic or Holocene material.

4.7   Organic evolution model

Evolutionary changes are regarded as essentially gradual over the time-scale concerned, but they were clearly intermittently catastrophic in restricted areas for specific durations of time. For the land areas of the Mesozoic world, the most simple possible model calls for natural selection (or, more accurately, natural survival) resulting from geologic physical changes of both a horizontal and vertical tectonic nature and changes of climate resulting from that movement or caused by external effects. Although these changes are regarded as having been gradual over long time intervals, there is no need to distinguish any other model, such as punctuated equilibrium, that may have applied locally on occasions. In view of the rather limited numbers of plant taxa recorded (Hughes 1976, chapter 5), and the probability that the whole upland land surface was by no means fully colonised at that time, it seems unlikely that strictly biological competition and control were significant in such ways as were clearly operative in Cenozoic times.

As is discussed below, the direct influence of Mesozoic animals on plant life was probably limited and far less than our specialised Holocene experience of the integrated biological world may suggest. For example, plant root systems themselves, and the soils they responded to, would have been only part way towards the diversity we now know. Although physical and chemical laws will not have changed, all chemical compositions and concentrations evolve and with them the nature of all environments. It is difficult to allow in thought for so much change, but it is important not to expect 'uniformity' through geologic time of any natural scenario, such as mangrove sea-margins. Just as no part of the earth's crust may be thought of as constant in composition or in location, no biological interface should be regarded as surviving unchanged; evolution is universal and continuous.

4.8   Mesozoic 'Pollenifera'

All Mesozoic seed-plants are here referred to neutrally as 'Pollenifera', so that the term gymnosperm may be removed from contention for most of that era. Most Mesozoic seed-plants were clearly not angiospermous, but it has also proved unhelpful to refer to them as 'gymnospermous' and then to classify them into groups typified by the various impoverished relicts that are the 700 or so species now living. This blinkered view from the Holocene, classes everything

Mesozoic as 'coniferophytes' or cycadophytes or just possibly gnetalean, all on very low quality evidence. In fact Mesozoic seed-plants, as consistently pointed out by T. M. Harris throughout his career, were very much more diverse than are gymnosperms now; nothing has been gained and much inspiration lost by unimaginatively using these classes, despite the warning by Arnold (1948).

4.9   Angiosperms are polyphyletic

The angiosperms are assumed to be of polyphyletic origin until proved otherwise. The main concept of the angiosperms as a monophyletic group depended on the belief that the curious and interesting double fertilisation and eight-cell female gametophyte was constant in detail throughout 250 000 species and that such a structure could not have evolved twice even in 100 million years. It is extraordinary that such an obstructive piece of conjecture could have been so little challenged and thus could have remained as dogma for a century. It is time that alternative views should in their turn be tested to the full, regardless of which view may prevail in the end.

4.10   Period classifications

Classification of plant fossils are erected here only for specified geologic periods, to avoid inclusion of any (irrelevant) data from later periods. As discussed by Hughes (1989, chapter 10), when classification proves to be necessary because of the quantity of material to be recorded, separate independent classifying schemes are set up for each successive geologic period or for lesser durations of time; these schemes include only material recorded from the period concerned. Higher taxa are set up only when the quantities of base taxa involved merit this; no hierarchical terms derived from later periods of time are admitted.

4.11   Unique climatic maximum

All geologic periods are unique in evolution and the Cretaceous period represented the climax of the main recorded anti-glacial phase of earth's history. The earth was presumably warmer, although perhaps not greatly so, was virtually ice free and therefore experienced an unusually high sea-level under eustatic control. The latitudinal gradient of temperature (Barron 1983, Spicer *et al.* 1993) was less developed than in a glacial period such as the Quaternary or the Carboniferous–Permian, at least as indicated by Late Jurassic and Early Cretaceous floral and faunal distributions.

4.12   The principles and the problem

As several of the principles represent a direct reversal of current customary lines of thought in distinct fields, it is desirable that they be challenged and questioned as necessary before we accept this perhaps unfamiliar alternative framework. The difficulty, however, of conducting such questioning on this topic is well illustrated by the rather long and emotional reply of Crane (1984) to the timely philosophical probing of Mabberley (1984 [same journal]). The solution of the problem of angiosperm origin is a matter for the long term because there is so much recording and investigation of fossils yet to be done before even the possibilities already envisaged become exhausted. The necessary large-scale data handling will be rewarding only if the very large quantities of data on microfossils yet to be collected are clean enough and unambiguous. The eventual solution of the problem will become possible if all the old confining dogma is challenged and tested.

5 Stratigraphic framework

5.1 Statement of stratigraphic age

Of more lasting importance than the necessary details of rock formations and zonal schemes is the need to accompany every mention of an observation record of fossils with a statement of estimated stratigraphic age. If precision is not possible, the statement should give a bracket of age limits, however wide this may have to be; even a two-period bracket from early Jurassic to end Cretaceous has some value, and can always subsequently be refined (narrowed) progressively as further information is obtained. It is no exaggeration to say that fossils, or theories based on them, without such a statement to accompany them at every mention, are worthless for any study of phylogeny or of evolution.

5.2 Global stratigraphic time-scale

For several decades geologists have been steadily developing a set of agreed subdivisions of geologic time applicable world-wide; as for example Arkell (1956), who standardised the divisions of the Jurassic period. More recently the Stratigraphy Commission of the International Union of Geological Sciences (SC-IUGS) has been overseeing and stimulating the creation of precise definitions for all of the boundaries concerned between the divisions of such a scale (Harland 1992); the required formal designation is now of a Global Boundary Stratotype Section and Point (GSSP) of Cowie *et al.* (1986), taken in a single carefully selected rock succession with accessibility and good outcrop, adequate sediment thickness and continuity, and rich fossil content. Relatively few such points have yet been agreed finally, but current usage of unregularised points is sufficiently important to be the major factor in future agreements, and large changes are thus unlikely. None of the beginnings of the three Mesozoic periods is yet formally agreed; the ends of periods are not separately registered but are each taken at the beginning of the next later period. A recent state of the whole scale and its nomenclature is given by Harland *et al.* (1990). The successively smaller-scale divisions are: era (e.g. Mesozoic), period (Jurassic), epoch (Dogger = Mid-Jurassic; a division considered unnecessary by some), age (sometimes called stage) (e.g. Bajocian), chron (usually named on fossil content; duration frequently less than 1 Ma). In this work, the smallest division used is normally the 'age'. The beginning of the first age in a period is always taken at the point for the beginning of the period (e.g. Cretaceous and Berriasian beginnings will be defined at the same point). The divisions have thus been progressively formalised over many years, and their general pattern is derived from knowledge of their fossil content and succession in the field (Rawson *et al*. 1978, Cope *et al*. 1980).

5.3   Geochronometric scale

The scale in millions of years is sometimes considered simpler to memorise and thus more attractive for use. In practice, however, there are relatively few and irregularly distributed tie-points where suitable rocks occur for radiometric determination, and a chronometric scale derived from these facts is used for calibration of the main stratigraphic scale. For example, the Cretaceous Cenomanian age beginning (= end Albian age) is estimated to have occurred at 97 Ma, although there is an error-function depending on different factors in the calculation for each individual boundary case (see Harland *et al*. 1990, p. 200 *ibid*., and explanation p. 197). The relevant part of the complete scale in use here, of Mesozoic ages and their geochronometric calibration, is given in Fig. 5.1.

5.4   Other scales

Much has been written about alternative scales based on paleomagnetic reversals, on sea-level changes and on numerous other phenomena, but none of these is independent of fossil evidence. The evolving floras and faunas of the world provided unique sequence evidence in which no set of observations is ever fully repeated; one paleomagnetic reversal phase depends on the same physical measurements in rock samples as those in any other reversal phase, and they can be distinguished only through the additional evidence of the fossil sequence. There is thus only one Stratigraphic Scale (see above), while any other data such as sea-level changes may help merely to characterise its divisions.

5.5   Stratigraphic correlation

Recording the detail of rock successions represents arduous fieldwork; selection (Fig. 5.2) of individual successions to serve as boundary stratotypes of scale divisions involves wide experience and world-wide co-operation. The most difficult stratigraphic exercise requiring all skills is in relating rock successions (which will normally differ greatly in detail) from different areas. This is done principally by using fossils, but it is always necessary to qualify any correlation with a statement of probability of accuracy (see Hughes 1989, chapter 14). The

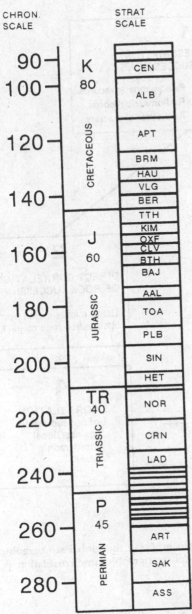

CHRON. SCALE

STRAT SCALE

Fig. 5.1. Stratigraphic scale of periods and ages from Permian to Mid-Cretaceous, with chronometric calibration; data from Harland *et al.* (1990). Abbreviations from the same source; names of very short ages from Permian Kungurian to Triassic Anisian omitted, also End-Triassic Rhaetian and Cretaceous Turonian.

Abbreviations used for ages in all similar illustrations in this book are as follows: CON, Coniacian; CEN, Cenomanian; ALB, Albian, APT, Aptian; BRM, Barremian; HAU, Hauterivian, VLG, Valanginian; BER, Berriasian; TTH, Tithonian; KIM, Kimmeridgian; OXF, Oxfordian; CLV, Callovian; BTH, Bathonian, BAJ, Bajocian; AAL, Aalenian; TOA, Toarcian; PLB, Pliensbachian; SIN, Sinemurian; HET, Hettangian; NOR, Norian; CRN, Carnian; LAD, Ladinian; ANS, Anisian; SPA, Spathian; NML, Nammalian; GRI, Griesbachian; CHX, Changxingian; LGT, Longtanian; CAP, Capitanian; WOR, Wordian; UFI, Ufimian; KUN, Kungurian; ART, Artinskian; SAK, Sakmarian; ASS, Asselian.

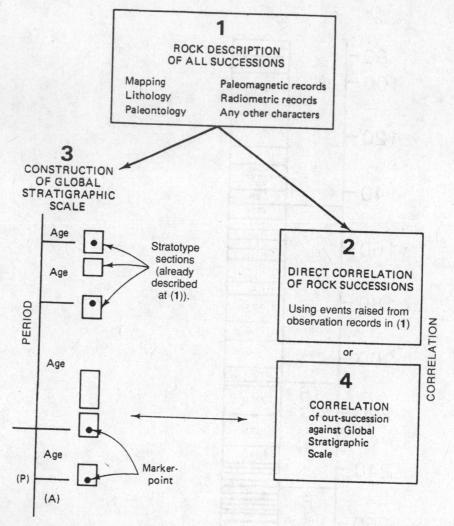

Fig. 5.2. The main activities concerned in erecting the global stratigraphic framework, leading to the possibility of successful stratigraphic time correlation. p, period; A, age (from Hughes 1989).

difficulty of doing so always increases with paleodistance from a Stratigraphic Scale Point reference, and particularly across a paleolatitude climatic gradient.

5.6   Application to individual observations of fossils

The recording of the occurrence of relevant plant fossils, particularly of those suspected of evolutionary significance in the search for angiosperm origins, requires a positive intention to distinguish inequalities and to draw attention to differences. This has to be unashamedly opposite to the more usual employment of other commonly occurring fossils in stratigraphic correlation in which the demonstration of adequate similarity for comparison of the time of occurrence of the organism normally prevails as a purpose. As explained by Hughes (1989, chapter 4), it is essential for effective evolutionary studies to record and file data about each actual observation of fossils, particularly the small differences from other records of supposedly the 'same taxon'. Dead-end approximated identification by an observer of unknown competence and with knowledge dating from several years earlier, actually prevents necessary investigation in this context, and represents a confusion of intentions.

5.7   Intention of employing Period Classifications

Just as the unqualified identification of specimens to taxon can obstruct, so also the fitting of that taxon into a classificatory hierarchy indicating that it is a 'pteridosperm' or a 'cycadophyte' provides unwanted and usually misleading association of other data when applied to a Mesozoic plant fossil of which probably only one organ is known. The solution adopted here (see Hughes 1989, chapter 10) is to assemble, and classify if necessary, only data from observations made in a stated period of geologic time; no reference is permitted to any organism or fossil derived from a later period, on the grounds of the irrelevance of a plant not yet evolved; similarly no serious thought is given to the future, when living material is being classified.

5.8   Designation of the selected periods for classifications

The choice of definition on which to base a selected 'period' is as follows:

(a) An easily remembered round figure in million years before present (Ma).
(b) The end of a time-scale age (= beginning of the next one).
(c) A bed or horizon as a visible feature in rock from which a main collection of fossils has been taken.

The attractive objectivity of (c) is marred by the fact that it can refer to only one succession at one locality and for others is subject to the problems of correlation to the Stratigraphic Scale. The round figure (a) and the end of a time-scale age (b) rarely coincide and, as calibration is refined, the Ma figure will change within the scope of the error-function for that division, which may be of the same order as the duration of the division. It is also obviously desirable in such an arrangement only to designate Period Classifications as they become necessary or are required, and to leave scope for additions at a later date. Figure 5.3 illustrates the application of this selection to the well-known Yorkshire Jurassic fossil occurrences, by selection of option (a) regardless of the possibility that a reassessment might place the

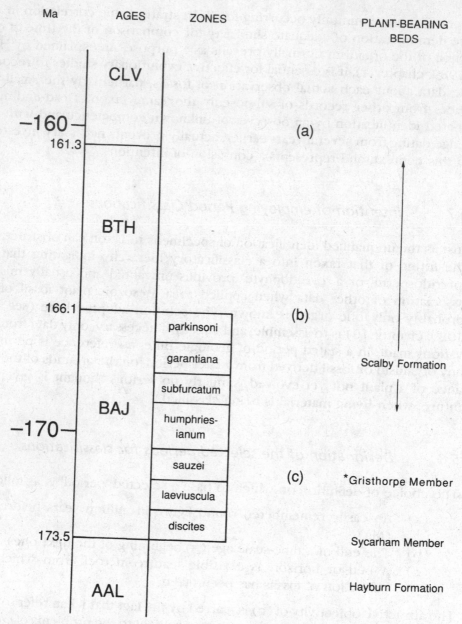

Fig. 5.3. The principle for defining the scope of a Period Classification. The Aalenian–Bajocian–Bathonian–Callovian sector of the Jurassic division of the Standard Stratigraphic Time-Scale is shown with estimated ages (from Harland *et al.* 1990). The Gristhorpe Member has yielded the most diverse and well-preserved flora, but much material has been collected from the other beds named. Choices (a), (b) and (c) are discussed in the text.

Bathonian/Callovian boundary after 160 Ma ago; it is also clear that principle rather than precision should govern such a choice (a), which is now followed in the other selected periods (see Chapter 14).

5.9   Post-Cretaceous stratigraphic divisions

Although there is no intention to discuss Cenozoic angiosperm development in this book about the origins, some reference is made in later chapters to divisions of Cenozoic time. The Cenozoic era refers to all time subsequent to the Mesozoic era and up to the present inclusive, embracing the old Tertiary and Quaternary. Tertiary subdivisions derive from the arrangement of Lyell as modified even in the last century: Paleocene, Eocene and Oligocene sometimes grouped as Paleogene, and Miocene, Pliocene sometimes grouped as Neogene, with a total duration of 64 million years. The Quaternary comprises the Pleistocene of 1.63 million years and the Holocene of 10 000 years up to the present. All individual age details are provided by Harland *et al.* (1990).

5.10   Note on paleocontinental maps

The cylindrical equidistant base maps are from a geologic database and software compiled by A. G. Smith and colleagues. The latitudinal error in the positions of the major continents is probably less than 5°. The position of all areas affected by an orogeny younger than a given map is uncertain, the uncertainty increasing with the age of the area affected. Continental shelves (e.g. in Smith *et al.* 1981) have been omitted for greater clarity, and 30° of longitude in the Pacific area have been cropped from most maps to aid presentation.

6 General Mesozoic palynologic evidence

6.1 The stratigraphic interval

The palynomorph record of pollen from approximately Triassic Anisian to Cretaceous Hauterivian time may appear to be relatively unpromising material to search for angiosperm origin evidence. So far it has only been generally explored by light microscopy in many useful but scattered reconnaissance papers, but these papers do no more than indicate the probable lines on which the next work can follow. For example, the non-saccate monosulcate pollen types are seldom described and developed because they are usually small; at low magnifications they appear to lack distinguishing characters and may well be overlooked or at best dismissed into one or two long-range taxa. These monosulcates are, however, believed to include pollen of many important megafossil plant groups of the Jurassic period, among which are the principal regular suspects as angiosperm ancestors.

6.2 Comparison of material

In order to emphasise the stratigraphic sequence of occurrence of fossils, every effort will be made to compare palynomorphs with their predecessors and never with their successors in later periods (following principle (5), see p. 21). Naturally, however, general observations on physical properties and probable function of pollen draw on studies of Holocene material to provide models, but without any taxonomic or systematic implications.

6.3 The group 'Pollenifera'

This group term (without rank) is deliberately used here for all the Paleozoic and Mesozoic seed-plants considered, with the purpose of eliminating the term 'gymnosperm' and all its connotations from this study. The properties of gymnosperms are all based on observations of living or Holocene whole plants. After one or two observations have been made on a Jurassic twig or leaf suggesting some resemblance of that organ to a part of a selected living conifer or ginkgo, it is inevitable that most or all of the other characters of the complete living plant become transferred in the mind and attributed to the Jurassic material.

Although some workers have considered this process justifiable or even desirable, others who genuinely try to separate the concepts find it understandably difficult to maintain such separation in a permissive literature (Arnold 1948). Consequently all classificatory terms such as conifer, cycad, ginkgo, gnetalean have here been removed from use in the Mesozoic and only names based entirely on fossil occurrences are employed (see also Hughes 1976, chapter 8). It is important for clear understanding and discussion that the characters of plant organs not yet known as fossils should remain as blanks in the knowledge matrix. Reconstructions of past forests are perfectly acceptable as imaginative exercises, but neither the database nor any postulated phylogeny should be contaminated or affected by such speculation. The term Pollenifera is also intended to include possible Mesozoic angiosperms until they are individually formally accepted into the otherwise Cenozoic and perhaps also Cretaceous angiosperm group. This procedure removes any difficulty from classifying the Barremian monosulcate pollen (see Chapter 9); a decision on this and other such problematic material can thus be deferred until adequate evidence has been obtained.

6.4 Interpretation of function in pollen

As restated clearly by Thanikaimoni (1986), the functions of the pollen exine of the aperture and of the main wall can be summarised as provision for germination and for the ion exchange associated, for protection of the cell and for harmomegathic changes of volume. There is no reason to regard Mesozoic pollen as differing in any of these functions, although the relative importance of any one function may have been unlike that in any living pollen. Much more is known about such properties as pollenkitt in living pollen than from fossils, but correspondingly until very recently great unappreciated damage was caused during processing such as acetolysis and staining of recent pollen. With fossils, oxidation of the exine to improve translucency has caused similar or worse damage in the past but as regular observation technique changes over to electron microscopy this potentially destructive part of the hitherto normal processing of samples can be and is being reduced greatly, or avoided altogether.

6.5 Mesozoic pollen in general

Accepting pollen grains as only one organ of a plant and therefore not to be considered as evolving independently, it is still possible to examine a succession of states of elaboration reached by pollen grains as a result of partial selection pressure operating naturally on this organ through several geologic periods. One view of Mesozoic pollen is that as a whole it appears to display relatively little change in or development of organisation through two geologic periods of over 100 million years up to Cretaceous Hauterivian time. Other organs or organisms, both plant and animal, do not appear to have been so development-ally limited in the same interval. Pollen of this kind, however, was constrained in organisation uniquely in its prime function of passing through a narrow tube micropyle to the ovule of all early 'Pollenifera'. It was also constrained geomet-rically from the early stages in the Carboniferous period when a distal pollen aperture was developed for germination, presumably freeing this elaborate func-tion from the production complication of parallelling the proximal laesura in the meiotic tetrad.

6.6   Symmetry

All of these late Paleozoic and Mesozoic pollen grains were distally either mon-oporate or monosulcate in basic organisation; monoporate grains with a single round (or possibly ulcerate) pore at the distal pole may presumably be consid-ered to be derived from a trilete miospore from a tetrahedral tetrad. Monosulc-ate grains with a single short sulcus or long linear aperture over the distal pole could have come from a monolete miospore of a tetragonal tetrad, but could equally represent a variant of the trilete spore from a tetrahedral tetrad. Other possibilities appear to have been rare enough to lack significance either in the Carboniferous palynofloras or in subsequent derivatives. Pollen grains without any apparent germinal aperture can usually be regarded as monoporate in func-tion at germination.

6.7   Sacci

Saccate grains were particularly diverse in Permian and Triassic time, and con-tinued later in less variety. All saccate grains are essentially of the two symmetry types mentioned above, and are thus simply variations of either monoporate or monosulcate organisation. Although some botanists have postulated other functions, it seems most likely that all sacci performed the single function of acting as 'water-wings' and were concerned only in upward flotation in a tem-porarily or permanently downwardly inclined micropyle; if the sacci served or serve any other purpose, there is no sign of such a development in any known angiosperm pollen and TEM scrutiny of living bisaccate pollen (Kedves and Pardutz 1992) has only begun. Sacci are only meaningfully positioned in an inflated grain, and as the development of sacci long preceded entomophily, the micropylar fluid can have been the only appropriate agent of such inflation. The probable haustorial function of the pollen tube in some cases does not alter this conclusion. The usual slightly distal placement of the sacci is further evidence of a function of directing the attitude or orientation of the germinal aperture.

Monosaccates with apparently equatorial sacci probably relied on differential thickening of the exine at the two poles to achieve a similar orientation effect.

6.8   'Striate' saccate pollen

In late Permian and early Triassic time, a majority of all bisaccates were also taeniate (although obversely termed as striate) with numerous parallel straps of exine thickening across the proximal area, essentially between the roots of the sacci. The functions of such taeniae have been regarded as harmomegathic, and this is certainly acceptable if it is to be taken to include prevention of any damaging collapse in the dry pollen grain at the time of distribution. The lack of such 'striate' pollen in Jurassic and later times suggests that perhaps more subtle and less materially expensive exine developments could have fulfilled this function by Jurassic time; a possible minor but interesting exception is *Decussosporites* Brenner 1963 from Potomac Group Zone I of Early Aptian age. The interest in these prominent features has even resulted in persistent claims that Permian *Vittatina* and Triassic *Equisetosporites* have affinities with the Gnetales of which some Holocene members show similar features; the complete lack of megafossil evidence of the three extant gnetalean genera before late Cenozoic time suggests that such unsubstantiated claims should be disregarded and treated as harmful to clear, unbiased reasoning (Hughes 1992) but see also Section 8.31 below.

6.9   Monosulcate pollen

Most Jurassic and Early Cretaceous pollen, at least throughout the northern hemisphere, was monosulcate with all the bisaccates regarded as comprising only a slightly modified version of the monosulcate. This included pollen of all plants of the linearphylls (see Hughes 1976), Bennettitales, Nilssoniales, 'Ginkgoales', and some others. This formed a clear majority of all Pollenifera (seed-plants) that were present in the megafossil record; in very many counts and in virtually all so-called paleoecologic summaries no further distinction is made beyond bisaccates and simple monosulcates. The reasons for not distinguishing further are the low magnifications and the great difficulties in achieving usable and repeatable orientations of dispersed bisaccate pollen. This sector of mid-Mesozoic palynofloras has remained unsatisfying in study, and has consequently been relatively neglected.

6.10   Monoporate pollen

Seldom numerically dominant in Jurassic times even on a local basis, there were simple forms such as *Exesipollenites*, together with some monosaccates such

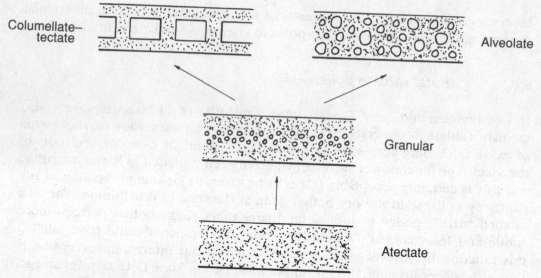

Fig. 6.1. The extension of the term tectate as applied to the outer exine of pollen, from the commonly understood columellate–tectate of many angiosperms, to include granular and alveolate ektexines, which are found in other groups of Pollenifera; the relationships shown are simplistically conjectural. (After Traverse 1988.)

as *Callialasporites* and polysaccates such as *Cerebripollenites*. Additionally in certain paleolatitudes the remarkable and distinctive *Classopollis* group (see below) was unusually common; as far as is known the brachyphyll megafossil plants (Hughes 1976) provided much of this pollen.

6.11   Tectate pollen

For many workers, tectate pollen is an angiospermid feature in which an outer ectexine surface tectum layer is supported on numerous short radial structures known as columellae, the whole making a roof-like cover over the main exine pollen wall, hence the name; the tectum is in many cases incomplete (semitectate) so that the intercolumellate spaces are in part apparently open to the exterior. The term tectum, however, has become downgraded by its wider use for any continuous outer layer, over a granular or otherwise discontinuous middle exine layer that may be described as alveolate in these circumstances (Fig. 6.1). These granular and alveolate exine structures are recorded in some Mesozoic Pollenifera; they may perhaps be regarded as an evolutionary step towards the columellate–tectate angiospermid pollen of Early Cretaceous time. Until it is substantiated by recorded successions of such pollen, it may be premature to suggest this because it is evidently not the case in the elaborate exine structure of the *Classopollis* group, which persisted independently into Late Cretaceous time.

It is important, however, to record that pollen wall 'experimentation' was a continuing Mesozoic feature despite the relatively unchanging outward appearance of the dominant monosulcate and bisaccate monosulcate pollen. 'Experimentation' would have concerned not so much the passage through the micropyle attuned to varying sculpture and external shape, but to the development of structural features within the exine that might ameliorate the effects of harmomegathic and other constraints.

6.12   Speculation on successive changes of function

A saccus is not purely a hollow space but often contains a complex internal mesh of supporting structural elements; this structure presumably facilitates the retention of some air in the saccus during dessication-collapse but is ready to adjust with air during the fluid inflation of the main pollen cavity on micropylar fluid immersion. The development of sacci over a considerable part of the grain surface is analogous to that of a tectum, providing some rigidity and strength; in some later bisaccates a reinforced exine structure known as a 'cappa' extended beyond the sacci and in some cases over the whole proximal surface, perhaps replacing the function of the Triassic taeniae. For times earlier than the main Cenozoic development of specific entomophily, a structural purpose appears to be the only reasonable explanation for such elaborations, but subsequently many such features could have been transformed to store insect attractants, wetting agents and compatibility recognition features. Such important changes could have taken place steadily in Mesozoic pollen without any striking outward signs of change. Cretaceous columellate–tectate monosulcates of Hauterivian–Aptian times (see also Chapter 9) could have represented a culmination of such trends of functional change in monosulcates at least as far as insect attractants for Coleoptera. In this case these Early Cretaceous pollen grains could have belonged to plants without either flowers or any other angiosperm attribute, but thus far there is no evidence either way.

6.13   Classopollis *group*

Members of this group are known from palynologic assemblages from mid-Triassic time to late Cretaceous. Megafossils are known from Late Triassic (*Cheirolepidium*) through to Early Cretaceous time (*Frenelopsis* etc.) but the two plants concerned were not similar and did not overlap ranges in the Jurassic. Although they may all come from the brachyphyll group, a variety of plants was obviously involved. Curiously, with the *Classopollis* pollen, much study time has been spent on legalistic nomenclature problems resulting from lost types and poor early descriptions and interpretations (see Traverse 1988), and rather little time on splitting up this large group stratigraphically into successive taxa distinguishable on both sculpture and structure criteria. Both Reyre (1970) and Medus (1977) from France have attempted this discrimination on French and North African material, respectively, but they have rarely been followed in other stratigraphic assessments. Considerable effort has gone instead into paleoenvironment interpretation (see e.g. Francis 1983 on the late Jurassic of southern England); in several other cases little regard has been paid to evolutionary development through a stratigraphic succession representing 100 million years. *Classopollis* is frequently, but by no means always, found in tetrads and it seems probable that with less vigorous routine palynologic preparation of samples many more such tetrad observations might have been made. The persistent tetrad is surely functional, and the resulting all-round presentation of a germinal pore suggests some topographic difference in the pollen reception area, either in or adjacent to the pollen chamber or on some simple stigmatic surface.

6.14   Late Triassic monosulcate pollen

Since Cornet (1977) wrote in a thesis for Pennsylvania State University about a diversity of Late Triassic tectate pollen, there has been revived interest in previous hints (Schulz 1967) of such occurrences. A few of his fossils were figured by Traverse (1988, p. 229) but Cornet (1989a) has now published descriptive records of a new group Crinopolles with several species. The material was originally compared with the known Cretaceous Barremian monosulcate tectates but the Crinopolles differ strongly in having a much greater grain size as well as in sculpture; they appear to be distinct and perhaps to reflect an independent Late Triassic radiation of plants also recorded in separate megafossils (see Chapter 8). They do not appear to have immediate successors in the next 75 million years and are perhaps best regarded as further evidence for much greater diversity of Mesozoic Pollenifera than has been previously admitted. This provides a further chance of ultimately tracing angiosperm ancestors through successive occurrences in Mesozoic rocks.

6.15   New Jurassic monosulcate evidence

Cornet and Habib (1992) reported crotonoid monosulcate tectates from Jurassic rocks of undisputed Oxfordian age on the Normandy coast of France. Unfortunately this exciting find has a number of very odd features and calls out for a quick repeat confirmatory observation as the rock succession concerned is from a straightforward coastal exposure. The sample was collected by others for a different purpose and was given to one of the authors and processed fifteen years ago; the current surviving fragment of sample is apparently too small to process further. The main find consisted of thirty-three specimens of a new crotonoid species *Stellatopollis pocockii*, which had been single mounted (glycerine jelly) including twelve specimens in clumps of four; nine specimens including the photographed holotype had apparently disintegrated by June 1991 and several others were greatly distorted by swelling. A paratype was mentioned but not figured. The average grain size of measurable specimens was an enormous 92 µm. Found with the *Stellatopollis* were three grains identified to *Multimarginites* Germerad *et al.* 1968 of diameter 114 µm, attributed to the Holocene (and Tertiary) family Acanthaceae; again, of the three, one grain disintegrated and another swelled up. All the regular pollen, spores, and dinoflagellates of the sample were undamaged and of normal size and appearance, but they had not been single-mounted. *Stellatopollis* was recorded as 0.4% in a count of 235. Several other dated samples from adjacent rocks were processed similarly and provided regular palynomorphs but no angiosperm-like pollen. The authors are to be congratulated on publishing the details of this surprise record after obviously experiencing very great difficulty; they or others should now repeat this important observation quickly before it becomes incorporated unchecked in textbooks and SEM study should be included. In the meantime, this record should be treated as 'pending' in the usage of Muller (1981), and therefore not built into any hypothesis.

6.16 Manipulation of pollen character matrices

Doyle (1988) provided for pollen the kind of review and cladistic analysis given in many other papers for all characters of plant groups. As usual this is presented as a copse of so-called parsimonious 'trees', with embarrassing confidence verging on religious orthodoxy. The saccate condition is treated as a single-origin derived state without any stratigraphic or other justification, the taeniae of Permian pollen are taken with the 'striations' of *Ephedra*, and the anthophytes are rigidly monophyletic. The database of plant higher taxa in use is traditionally organised and unsurprisingly the results follow this. When such studies are made entirely within selected geologic periods using only characters recorded therein, the method may well be truly tested. Doyle (1988, p. 14) referred to 'some cladists rejecting the use of stratigraphic evidence', but failed to condemn them; it is difficult to detect any progress in this deliberate chewing and rechewing of the same data, however numerical the methods employed.

6.17 Further study

Discrimination of different wall structures in virtually sculptureless monosulcates from Early Jurassic through Cretaceous Valanginian strata clearly requires minimal processing (without sieving), SEM and TEM study, and almost certainly SEM search to locate the probably very small specimens. This has not yet been attempted, although some strata from southern England have been searched during the hunt for the monosulcate tectates of Barremian and Hauterivian time and their entry into the succession. Such study will almost certainly reveal much more variety and evolutionary development than is now documented, as has already been done for some Late Triassic occurrences by Zavada (1990).

7 Integration of Mesozoic floras and faunas

7.1   Independent assessment

Over such a long period of time as the Mesozoic era not only the animal and plant components but also their relationships must have evolved considerably. The very closely integrated land plants and animals of Holocene and late Cenozoic time that comprise usual observations now are a very long-evolved and so far short-lived phenomenon; this is so different as to provide little guidance about conditions in say Triassic and Jurassic time, or even at 100 Ma when angiosperm arrival was certain. An entirely separate assessment is necessary.

7.2   The dinosaurs

Clearly the dominant land animals in Mesozoic time must have had a relationship with the more advanced land plants (Pollenifera), although it was not necessarily a direct connection by grazing, nor was it likely to have been a static one through so much time and change overall. In some respects the dinosaurs are easier to interpret than the plants because they are believed not to have had any comparable living successors to confuse the issue. On the other hand, less attention has been paid to the dentition and to other details of the herbivores that at various times do not appear to have been adequate in numbers to match in any modern sense the scale of occurrence of the more attention-pulling carnivores (see e.g. Lambert 1989). Rather naturally also, much less is said of the smaller members of the dinosaur group, which must have been entirely integrated into the food-chains, as must have been lizards, other reptiles and even the small mammals that are less frequently mentioned, in the form of an important unavoidable interface with the Insecta of this time. A typical tantalisingly brief reference to gekkos of Barremian age from Spain was made by Kohring (1991).

7.3   Relationship of fauna to flora

All the land organisms evolved together and to some extent there must always have been some effects of the presence of a fauna on all the plants under investigation as possible ancestors of angiosperms, even if only in the details of the distribution of the animal droppings. The detailed descriptions of the animals have been provided elsewhere by vertebrate and entomologic specialists, who normally also offer their interpretations of probable life styles. These can be expected to be well based in the morphology of the fossils and are not in dispute, but interpretation frequently embraces supposed facts about other organisms such as the plants, the details of which may influence some of the numerous guessing situations in the reconstructions.

Opinions, therefore, that are known or suspected to be different from those of some faunal experts are offered here in the spirit of testing. With different types of both insect and vertebrate it is suggested that, from their first appearance on land, there was in most cases a longish period in which they co-existed alongside plant life without, or with very little of, the close mutual dependence that later became strongly evident in Cenozoic time. The first flying insects are believed to have been from Carboniferous Namurian time, long after the early Devonian Rhynie chert was formed and included much earlier apparently flightless insects. Carboniferous and Permian insects, although they did not neglect the special opportunities provided by some concentrated plant reproductive structures such as pteridosperm ovules and even spores, were probably not involved in feeding on other *living* plant structures; the animal food-chain started with plant debris (trash), notwithstanding occasional earlier evidence of 'bitten' leaves or stems. This lack of integration apparently applied to early land vertebrates, which were all carnivorous in some way as judged by their dentition.

7.4   Continuity from sea to land

Although plants and some animals lived on land in early Devonian time (400 Ma), interrelationships appear to have been analogous to those in the then adjacent Continental Shelf sea from which they had all recently come; in some respects they remained in parallel until Late Jurassic time (150 Ma). Additionally shelf sea conditions must also themselves be considered as evolving, in that for example sea-floor sediment exploitation by infaunal animals such as bivalve molluscs and echinoids changed fundamentally in earliest Mesozoic time, the following appreciation may be regarded as sufficiently generalised to stand for Silurian–Devonian time. The primary food-chain base was in the plankton, which included unlimited comminuted debris of all kinds of organisms; filter-feeders on the plankton such as bivalves, brachiopods, echinoderms and the like were predated by cephalopods and others back to debris; above this came arthropods, larger cephalopods and fish vertebrates in various successive size grades but all predatorial. Such large algae as existed, with either calcareous or organic skeletons, may have been subject to the molluscan radula or more probably the simple physical processes of the water-body in their passage back to debris.

In the early occupation of the land, all animals were predators of one size or another down to but excluding those feeding directly on comminuted debris, which were apparently early apterygote insects, perhaps largely aided by fungi. There were no general herbivores (either vertebrate or insect) in the Coal-Measures forests of the Carboniferous, which probably existed mostly in pristine condition as indicated by the undamaged plant content of many coal-balls in widely separated areas; some minor exceptions can be found but this was apparently the normal condition. Permian and most Triassic vertebrates were carnivorous, and so probably also were a high proportion of insects, with their attendant arachnids. When reptiles evolved back into the sea as ichthyosaurs and plesiosaurs, and into the air as birds or pterosaurs (of which some may have taken plankton directly), these animals were also carnivorous.

Reproduction by all marine plants and animals alike in the sea was presumably for the most part a matter of release of vast quantities of eggs or larvae or equivalents. Land plants originally dispersed spores in this same way but were soon up against physically caused losses attributable to a less effective distributing medium, in addition to normal recycling by predation. Carboniferous and Permian times record all the early stages of devices that appear to make plant reproduction more certain, but the influence of and the parallel with earlier marine conditions was clear. A specially modified type of life pyramid concept is required for this early (but not to be called abnormal) time.

7.5   Permian insect life

In late Carboniferous and Permian times the earliest winged insects flourished and achieved great size, although they were mainly exopterygote groups with larval or nymph stages essentially like the adults and with no pupa. Many of the groups identified (from adult wings alone) have living representatives (Carpenter 1992), such as the predatory Odonata (dragonflies), but in other cases the living forms are less important overall among insects now than were their Late Paleozoic members in their time, particularly the Dictyoptera (cockroaches). The many Permian insect discoveries in the former USSR and in the USA have been placed in a number of extinct groups, such as the Palaeodictyoptera, the Megasecoptera, the Paraplecoptera and others, because the wing details unsurprisingly differed from those of the living exopterygotes.

With such incomplete knowledge of the whole insect life-cycle in those times, it is difficult to go further than this, but at least the temptation to classify the fossils into the nearest living group has been resisted by paleoentomologists studying the Permian. Overall the picture of these Permian insects consists of a number of predators and some that may also have adopted freshwater aquatic life in part, with a background of insects like Orthoptera (grasshoppers and locusts) and numerous others living among, and presumably also on, the fallen but not yet dispersed plant material (trash). The all-important exceptions were the Hemiptera (plant-bugs), with earliest verifiable fossils in late Permian time; these insects directly attacked living plants with their piercing mouth-parts, which could reach through cuticle to the phloem tissue in leaves and stems without directly destroying many cell walls. The great thickness and durability of most Mesozoic Pollenifera cuticle, which now so benefits study by paleobotanists, may have become a defence against these Hemiptera. Apart from some probably opportunistic raiding of Carboniferous ovules as indicated by elaborate plant defences of glandular hairs and very long micropyles in certain pteridosperms (*Lyginopteris, Physostoma, Salpingostoma*), there had been no direct systematic assault on green plants until this time (150 million years after land-plants arose).

Also in Late Permian rocks were fossils identified as Coleoptera (beetles), the first Endopterygota, presumably with specialised larvae and resting stages (pupae); these beetles had thick fossilisable elytra (protective wing cases) and, although differing in detail from members of living families, were clearly the forerunners of the most successful group of organisms on earth, with well over 300 000 living species, more than a quarter of all animals living now (Carpenter 1992). These Permian beetles do not appear to have attacked plants and probably lived in the fallen debris or trash (including spores) or on dung; some may have been freshwater aquatics like various living forms; the change, however, to the endopterygote life-style implied the presence of unfossilisable larvae that may have been involved in destruction of living plants by gross feeding. There is as yet no way of knowing about these larvae; much later endopterygote larvae of the Lepidoptera (butterflies and moths) of Cretaceous age onwards did and do cause obvious destruction of living plant tissue, but this damage has not been widely reported from observations on Permian plants and may well not have been a character of any early Coleoptera or their larvae. Thus it appears that, with the exception of the specialised Hemiptera, insect contribution to the food-chain in Permian time was still primarily of the products of dead rather than of living plant material, still a continuation of the Continental Shelf marine scenario of earlier times.

7.6   Triassic vertebrates and plants

The early Triassic existence of the supercontinent Pangaea probably aided world-wide distribution, but of the relatively small animals at this early stage. Although radiation of a variety of reptiles was swift, the size of domestic cattle was reached only in Late Triassic time, and that by carnivores (Lambert 1989). It has been suggested that these carnivores then gave rise to some plant-eaters but such an event supposes a very unusual adjustment to a strangely organised food-chain or pyramid. This different succession points to a lack of integration at this stage of the members of quite separate invasions of the land (see also Wing and Tiffney 1987).

The presence of plant-eaters requires some plants that were edible to them; although literature sources for vertebrates refer to ferns, horsetails, cycads, cycadeoids and conifers, the names amount only to comprehensive journalese. Several of the more primitive fern groups were represented at the time, and undoubtedly there were horsetails, but it is difficult to imagine any of these as a significant food source in Triassic time any more than they are now. Cycads, in the sense of the few relict extant genera around *Cycas* and *Zamia*, did not exist, although a few interesting fossil plants held perhaps one or two characters in common with them; cycadeoids were absent except as a misnaming of the Bennettitales *Williamsonia* group, of which there were few well-documented examples at this time. The low crown-of-leaves growth mode of these plants represented vulnerability to grazing, probably balanced by separate defence mechanisms. There were many unattributed seed-plants, but nothing resembling living conifers in detail. All the seed-plants mentioned had unusually thick cuticles, which have been widely prepared and studied; many had very small individual leaves and none of them would have offered anything easily digestible to a novice herbivore, even if possible defensive alkaloids are not considered. Although such a negative discussion is clearly inadequate, and bryophytes, algae and fungi may have played a part, there remains a conceptual obstacle.

7.7   *Jurassic dinosaurs and plants*

There were many more dinosaurs, the majority by then being perhaps herbivores, although these were not necessarily large. The plants differed greatly in detail and there was probably more variety than in Triassic time, but overall there was no widespread rapid innovation visible in the megafossil elements of Jurassic floras. The climate was believed to have become steadily globally warmer (Frakes *et al.* 1992); this was probably expressed (Sloan and Barron 1990) in a more gentle latitudinal temperature gradient but also a new monsoonal system would have resulted in changed arid regions. Surprisingly there have been few dinosaur finds in paleolatitudes greater than 35° Jurassic N, and the climate changes affecting the organisms in lower paleolatitudes might well have been negligible. It is tempting to assume that large herbivorous animals were direct gross feeders on larger plants such as trees and shrubs, but the conceptual difficulty of detail mentioned above remains. Most interesting are the large sauropods concentrated in Late Jurassic time; six groups of these, the Cetiosaurs, Brachiosaurs, Camarasaurs, Euhelopodids, Titanosaurs, and Diplodocids had pillar-like fore- and hindlimbs, long to very long necks and very small heads with nostrils high. The teeth are described as peg-like in two of these groups, as flat spoon-like in two and as chisel-shaped in two. In *Brachiosaurus* the teeth are described as showing wear but perhaps the converse would be more surprising in an organism supposedly capable of living for a hundred years.

However these animals may eventually be classified, Late Jurassic time was the principal period of radiation of these large dinosaurs and their allies, land herbivores of all sizes and carnivores to balance, birds and pterosaurs in the air, and plesiosaurs and ichthyosaurs in the sea. This may well have taken place in relative temporal stability of climate and non-innovation of any large scale in land vegetation. In the Early Jurassic, however, the existing herbivores were then all relatively small, comprising Ornithischia such as *Fabrosaurus* and *Heterodontosaurus* of about 1 m length from South Africa, displaying cutting teeth and grinding teeth, and armoured, slightly larger, forms such as *Scelidosaurus* and *Scutellosaurus* from the USA; the Pro-sauropods such as *Anchisaurus* and *Yunnanosaurus* of about 2 m length from the northern paleohemisphere had shredding ridge-teeth and chisel-like self-sharpening teeth, respectively. This all suggests a logical employment of the available early Jurassic land vegetation, however tough it might have been to gather and/or to digest, but entirely on a small scale. By Mid-Jurassic time, the Ornithischia were represented by 4 m *Dryosaurus* in North America and Europe, which had developed a beak and grinding teeth, perhaps leading on to Late Jurassic *Camptosaurus*, described as a bulky 6 m beast also from North America and Europe. The Late Jurassic Ornithischia from the USA included spectacularly armoured herbivores such as 6 m long *Nodosaurus* and 10 m *Stegosaurus* and allies with a clearly developed beak and no teeth at all, suggesting herbivores with strong defences rather than herbivores employing sheer numbers to flourish and survive. By contrast, the earlier unexceptional Pro-sauropod herbivores in Early to Mid-Jurassic time radiated to the six or more giant sauropod groups listed above that flourished briefly in Late Jurassic Morrison Formation (Dodson *et al*. 1980) and equivalent times, with some surviving in less dramatic form into and even through Cretaceous time. These very large sauropods have been advertised (Bakker 1987 and others) as high browsers on large trees with details of their limbs and pelvic girdles to support this possibility. This view was greatly extended by Coe *et al*. (1987) in a speculative work, although for various circumstantial reasons the high browser idea appears unlikely to be correct:

(a) Ornithischia at this time or soon after were successfully 'developing' (selecting) a much more suitable high browser in the form of Early Cretaceous *Iguanodon* and successors, with a beak and appropriate grinding teeth beyond a gap and light fore limbs associated with a different stance.

(b) Sauropods of this size would have been better suited to elephant-like behaviour in simply knocking down what they needed, but not with such very long necks and small heads.

(c) Most of the teeth mentioned (see above) were unsuitable for collecting tough vegetation but could well have aided raking and gathering of soft material.

(d) The much simpler explanation was of a quadripedal stance and horizontal grazing over shallow water of water-fern and similar floating material, with the long neck to provide adequate range, and the high nostrils and eyes to aid collecting.

(e)   It so happens that megaspores, probably mainly of such water-ferns, reach an interesting peak of diversity and occurrence at Late Jurassic and earliest Cretaceous time. Thus gross plant-feeding of this specialised type seems a more likely explanation, although it is no longer paralleled by any extant process; it is even possible that general water-level transgression was temporarily stalled at this time, giving widespread but short-lived appropriate conditions.

Natural selection (or survival) is most thorough and does not normally provide for unlikely although just feasible alternatives such as high browsing.

7.8   Small dinosaurs

Perhaps the smallest known is *Compsognathus* of European late Jurassic age, only about 60 cm long; it is described, however, as a flesh-eater with sharp teeth, which suggests that possibly the insects ate plants (or plant trash) in the food-chain and small dinosaurs came next above them. *Hypsilophodon* of the European and North American early Cretaceous is slightly larger at 1.5 m length and is described as a plant-eater with a horny beak and self-sharpening teeth, but by this time larger ornithopods also had these characters. Chicken-sized hypsilophodonts (Rich and Rich 1989) named *Leaellynasaura* have recently been mentioned from the Cape Otway Formation (Early Cretaceous) of Victoria, Australia (Douglas 1969, 1973). It is still difficult to tell whether there really were rather few of such animals, or whether taphonomy (see Spicer 1989, 1991), fossilisation, and consequent attention, have been unfavourable.

7.9   Early Cretaceous dinosaurs

By mid-Cretaceous time, net-veined leaves, some flowers and much triaperturate pollen clearly signal the presence of angiosperms and thus quite suddenly a major innovation to the floras that should perhaps be reflected in the plant-eating vertebrates. This may be very loosely rather than accurately correlated with the evolution of hadrosaurs in North America, Europe and East Asia and their persistence through Late Cretaceous time. For the first time, gross feeding on plants may have become simple and fully rewarding, with Pachycephalosaurs and Ceratopsids developing in such great numbers that the best-known carnosaur *Tyrannosaurus* could comfortably top the pyramid. By Late Cretaceous time this vertebrate and plant integration could be said for the first time to approach a modern style.

The climatic downturn at the end of the Cretaceous period was apparently enough in Maastrichtian time to damage fatally this balance that had taken so long to achieve; almost certainly none of the dramatic catastrophes, which have been so enthusiastically presented and may well even have occurred on some scale, are necessary in explanation of the failure of dinosaurs and of certain other (marine) organisms at this time. There was no land plant failure at the K/T boundary (see Chapter 12); a few much publicised North American 'fern-spikes' probably represent no more than the anticipated effects of local or even regional fire damage or other devastation.

7.10   Mesozoic insect life

The variety of suitable sediments for preservation of insect fossils has always been a problem and is particularly the case in Triassic time, although from Late Triassic rocks are recorded the first Diptera (flies); they were early members of a second, very large, group of living insects, which acted mainly as predators or lived saprophytically, but not on living plant organs. In Late Jurassic and Early Cretaceous rocks there are some extensively documented insect occurrences, in Kara Tau (Kimmeridgian) in Russia, in the Solnhofen Lithographic Stone (Kimmeridgian) of Germany and in the Purbeck and Wealden formations (Berriasian to Hauterivian) of Britain; there are a number of others of less importance, and the whole accumulation of records provides a fair summary of occurrence of possibly then existing groups (Whalley 1985).

The most recent account is that of Jarzembowski (1984, 1991), who has now found a much richer fauna than expected in the English Wealden in paleo-latitude 40° Cretaceous N where locally he has recorded abundance. He has clear evidence of Odonata (dragonflies), Orthoptera (grasshoppers), Dictyop-tera (cockroaches), Hemiptera (plant-bugs), and even Isoptera (termites), which are known now only from tropical and subtropical climates. From the Endopterygota he has Neuroptera (lacewings), Mecoptera (scorpion flies), Diptera (true flies) and Trichoptera (caddis flies), all essentially predators; Col-eoptera (beetles) were common as expected and also early Hymenoptera, the most evolved group of endopterygotes, but these do not include the higher Aculeata (ants, bees and wasps). Parallel discoveries were recorded by Jell and Duncan (1986) from Australian Aptian rocks. Not significantly present (Rasnitsyn 1984) were the Lepidoptera (butterflies and moths), apparently yet to radiate (Carpenter 1992). The degree therefore of gross feeding on living plants (the effects of larvae) remains undetermined but perhaps it was much more likely to have been significant by Cretaceous time. Actually proved as plant-feeders were only the Hemiptera, which were first recorded in Permian time (see above) and which are best regarded as parasitic on plant hosts.

7.11   Pollenation by insects in Mesozoic time

From the variety of exine tectum sculpture to be seen in early Cretaceous angiosperm and probable-angiosperm pollen, it is reasonable to postulate the involvement of several types of insect, although very little is as yet known about any flowers involved. In Triassic and Jurassic time, although there were other Pollenifera with flowers such as *Sanmiguelia* and *Irania*, the Bennettitales provided the best-known and primarily large examples. These were mostly unisexual as represented by *Weltrichia* (male) and *Williamsonia* (female), although the much smaller and rarer (and possibly unrelated) *Williamsoniella* was bisexual.

In very latest Jurassic and earliest Cretaceous time the larger-flowered *Cycadeoidea* was bisexual and probably protandrous; it appears that right up to this time, as described by Crepet and Friis (1987), by Crepet *et al.* (1991) and by others, only the Coleoptera (beetles) were likely to have acted as pollen distributors, and that in a fairly rough-and-ready manner. The main radiation of Lepidoptera (moths and butterflies) was to take place later. Although ants, bees or social wasps of the Hymenoptera were absent, the more primitive Symphyta-Xyelidae (hover-flies), associated now with some extant conifers, were recorded by Krassilov and Rasnitsyn (1982) from the Early Cretaceous of Transbaikalia with *Alisporites* and '*Pinuspollenites*' (but not angiosperm pollen) in the intestines.

As suggested by Pellmeyer and Thieu (1986), it is likely that in earliest Cretaceous time the principal and only important insect pollinators were beetles and the attractant for them would have been fragrance without involving a flower, rather than visual stimulus. The latest insect occurrences of Jarzembowski (1984, 1991) from the Weald Clay of Barremian age still did not include Lepidoptera; small moths listed elsewhere (Shields 1988) from this period are not recorded from the Weald. Higher Hymenoptera were also missing, although the monosulcate columellate–tectate pollen had begun a significant presence by this time.

7.12   Between vertebrates and plants

The Insecta, and their probable constant arachnid predators, must have occupied the apparent void between reptiles and the Pollenifera. No other land animal candidate such as the gastropod mollusca or worms can have credibly competed. Unfortunately as well as fossilising poorly, insects also spend little of their life-cycle as imagines (adults); larval stages, as well as being of totally different appearance and longer duration, have so far yielded little to paleontologists. Additionally, fossil insects suffer, as do angiosperms, from overuse of extant data in their classification and in the interpretation of their fossils. Although venation detail of a single fossil insect wing, even when slightly anomalous, will do duty for recognition of recent whole families and other higher groups, this is perhaps an inevitable reflection of the sparse state of knowledge that could, after some breakthrough, improve later.

7.13　Conclusions on plant–animal integration

Plant material energy has been translated during Mesozoic time into animal energy at five different levels subject to the overriding difficulty of animals digesting cellulose. Level 1 consisted of fallen plant debris (trash) in which the decomposition of cellulose was achieved from Devonian time onwards by naturally occurring bacteria and fungi on the ground, with apterygote and some exopterygote insects as immediate beneficiaries beneath a large pyramid of different-sized predators.

Level 2 consisted probably initially of a few more powerful insects and later of some small vertebrates feeding on concentrated reproductive plant materials in pollen, ovules and seeds; this very selective feeding was by small herbivores needing high energy foodstuff to sustain a high metabolic rate (Wing and Tiffney 1987). Some plants appear to have 'responded' even by Carboniferous time by defensively scattering rather than concentrating reproductive structures throught their vegetative parts.

Level 3 consisted from Permian time of hemipteran parasitic attack on internal transport of plant construction materials in vascular tissue; it seems possible that progressive thickening of leaf and stem tissue and particularly of cuticles throughout the Mesozoic amounted to defence.

Level 4 consisted of gross feeding by herbivores that were large among their contemporaries and fed continously, with a low metabolic rate, on low energy plant material held in the capacious gut for long periods with microbial fermentation therein to extract energy from the cellulose. The plant material was freshwater, floating, and rapidly reproducing species filling environments later occupied by *Salvinia, Marsilia, Trapa* and others, and characterised in Triassic and Jurassic time by an increasing number of independently shed megaspores. It is often naively suggested that because many Triassic rocks of Europe and North America are red coloured that these areas were primarly arid, although it is less frequently mentioned that much of the sediment, some of it coarse, was water-laid. Aridity there was in certain areas, but these were local and/or seasonally affected and there was ample scope for the parallel existence of standing water with vegetation, as in the Okuvango district of Botswana now. This temporarily significant feeding level appears to have reached a climax by earliest Cretaceous time with the giant sauropods.

Level 5 consisted of gross feeding for the first time on true land vegetation, including shrubs and trees, and this is marked by the radiation of large ornithopods with beaks to break off appropriate sections of such tough plants as *Brachy-*

phyllum, Czekanowskia, and possibly Bennettitales, which had not been grazed seriously before Early Cretaceous time. The mid-Cretaceous radiation of angiosperms made the vegetation richer but did not otherwise change the pattern, which endured until the end-Cretaceous dinosaur failure. The succeeding higher level of integration of plants and animals followed after some hiatus in Paleocene and Eocene times, when animal size was restricted, and is the basis of present-day observations. Birds, which were to become important subsequently, were in Aptian time (Sanz and Buscalioni 1992), still apparently aquatic and relatively cumbersome types not far removed from Late Jurassic *Archaeopteryx*; there was clearly no direct connection with land plants. In later Cretaceous time it is still difficult to suggest any trend in animals that might have been directly dependent on the angiosperm arrival, the most far-reaching of changes in the plant world.

7.14   Need for more data

In this field, for the present, speculation has outrun knowledge. The most important need is for a new, more reliable, floral reconstruction, based perhaps on the Berriasian–Valanginian Wealden of the Hastings area of east Sussex. Sedimentologic, taphonomic, palynologic, mesofossil and some megafossil evidence can be raised but the interpretation requires new inspiration and a complete freedom from the use of organs not actually seen in fossils.

8 Mesozoic megafossils

8.1 Introduction

This chapter is devoted to a commentary on seed-plant megafossils of the Triassic and Jurassic periods through to Cretaceous Aptian time, as they are the fossils among which angiosperm ancestors are normally sought. Several megafossils or groups of megafossils have been favoured intermittently as angiosperm ancestors, but in all cases raised so far claims have been dismissed because of clearly persisting 'gymnospermous' features such as the ovular micropyles and saccate pollen in *Caytonia*. On this kind of basis no fossil plant already known can ever qualify, and that is the root of the present impasse.

8.2 Customary view

It has become the custom for paleobotanists in this century to regard all Mesozoic 'gymnospermous' fossils as readily ordered or classed into groups of pteridosperms, or cycadophytes, or 'coniferophytes' with very few unattributed oddments. In addition there has been a tendency for all of these plants to be claimed as ancestors of one or more of the living gymnosperms, despite the fact that most plant fossils are only of isolated organs providing very incomplete information on the whole plant. This has left only the most unusual-looking and the more probably extinct plant fossils to be considered at all as angiosperm ancestors, with the additional constraint that for the whole large group of angiosperms only one (perfect) ancestor was thought to be required in this supposedly simple case.

8.3 An alternative proposal

Any alternative view needs to be as direct as possible in separating the observed facts from all subsequent assumptions about the *unobserved* organs that enter by association of images resulting from classifications that are probably obsolete or are at least not built round the fossils alone. The main proposal is to use the group term Pollenifera for *all* Mesozoic seed-plants, at least up to and including

Hauterivian time, and if desirable up to Cretaceous Albian time; this group would include any angiosperms possibly present until they have been effectively designated as such and removed formally to their own group. Thus any plant fossil displaying an advanced character would remain in Pollenifera until more details became available and until a satisfactory evolutionary succession had been built up to follow from it. As a consequence of such usage the terms gymnosperm, cycad, conifer, ginkgo and gnetalean (or chlamydosperm), and any terms constructed from these words, will not be used here for any Mesozoic plant fossil. This implies an almost complete removal of the attribution of Mesozoic fossils to extant families (or other taxa in hierarchies), as follows:

(a)   The small Jurassic and Cretaceous group Nilssoniales may well have been ancestral to living cycads of the *Zamia* or *Cycas* lines or both but the term cycadophyte is not needed to announce this ahead of proof by evolutionary links that is far from complete; the Bennettitales and *Cycadeoidea* are so unlike the Nilssoniales (Harris 1969, p. 2) that possession of a pinnate leaf may prove to have been the only parallel character.

(b)   At first sight it may seem acceptable that many Jurassic plants are referred to as 'coniferophytes' on account of their small or narrow leaves and occasionally because of their cones, but most supposed northern hemisphere Jurassic and earlier records of such leaves show perplexing differences of cuticle or other details, or lack of associated reproductive structures (see e.g. Fisher and Watson 1983), which must separate them from all the living families; only in Early Cretaceous time (Alvin 1957) is attribution of cones to the Pinaceae and to the Taxodiaceae (alone) more arguable, although not necessarily helpful to understanding of the evolution of these extant groups. In the southern hemisphere, because of the past influence of the work of Florin, numerous fossils such as *Rissikea* have been included in the Podocarpaceae for lack of another family, despite there being no diagnosis (extant or fossil) for this very diverse family and despite the occurrence of some similar contemporaneous fossils in the northern hemisphere as well.

(c)   The family and genus names for the single species *Ginkgo biloba* should have no place outside Holocene time; the fan-shaped leaf with dichotomous venation is indeed known and common from Permian time onwards but reconstruction of the *Ginkgo* name and thus of the plant has no justification.

(d)   *Ephedra*, *Gnetum* and *Welwitschia* have no place in Mesozoic rocks on the present negligible fossil evidence (Hughes 1992).

(e)   Even a 'pteridosperm' group is superfluous in Mesozoic time because there is no general definition for such a group and any common ancestry from Paleozoic pteridosperms has yet to be demonstrated. In effect the proposal refers all Mesozoic seed-plants to their fossil generic names only, informally grouped into brachyphylls, linearphylls, pinnates and fan-leaves if required, and united in the one neutral reference to Pollenifera. This is deliberately iconoclastic in relation to classification, but in no way to the observational records, which would not be affected. The purpose is solely to filter the useful (records) from the confusing (attributions).

8.4   Apparent loss to interpretation

Without the attributions from traditional classification, it is certainly more difficult to make reconstructions of vegetation and to discuss the paleoecology of plants under these distant conditions. Forty-five years ago, Arnold (1948, p. 12) wrote about Mesozoic floras that 'obsolete classifications retard progress and become misleading'; he might have gone on to say that a constantly republished past plant or floral reconstruction becomes an obstacle to thought, and there is much to be said for indelible incorporation of authorship and date in any such reconstruction, as an editorial requirement. With these in place, it could then be argued that there need be no real loss involved when the reader has been fully alerted.

8.5   Evaluation of the megafossil records

On close investigation, most of the unusual Mesozoic fossil plants are dependent for interpretation on a very small number of specimens (sometimes only one) or remarkably good preservation at one locality, from which almost all knowledge concerning them is derived. The stratigraphic position of this one locality is a matter of chance of preservation and does not necessarily relate to any mean range of occurrence data. In many cases the far less good specimens usually 'identified' under the same name, but often from other localities, conceal probably other species or at least records about which so far little is known. With these constraints in mind, the principal angiosperm ancestor possibilities are discussed serially below in approximate order of main recorded or of acme occurrence, with as much precision as possible about stratigraphic occurrence. This particularly can lead to assessment of whether any evolutionary succession of any value can be usefully pieced together.

8.6   Principal occurrence data

More than twenty distinct taxa of Mesozoic plant fossils have been or are considered as candidates for possible angiosperm ancestry. Several other less well-known plants probably should be and perhaps in future will be included. Virtually all have been found in the thirty or more significant floras world-wide that are shown with their paleolatitudes at the time of their formation in Fig. 8.1; surprisingly few fall outside mid-latitudes for the period concerned. In each

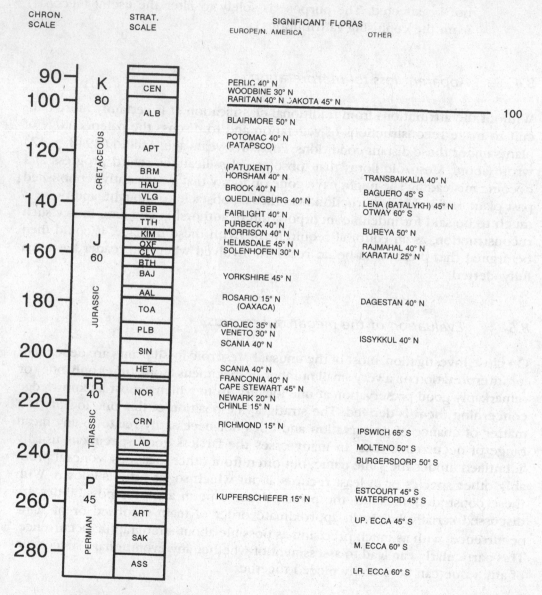

Fig. 8.1. The stratigraphic age of the principal (selected) Mesozoic localities and formations providing megafossil floras concerned in the discussion of angiosperm origins. UP., Upper; M., Middle; LR., Lower. The stratigraphic scale, age/stage abbreviations and chronometric scale calibrations are from Harland *et al.* (1990). Approximate paleolatitudes for the time of deposition of each flora are taken from Smith *et al.* (1981, updated 1992 by A. G. Smith personal communication).

individual taxon case, an estimate is offered for both geographic and time ranges, but these depend heavily on interpretation and on taxonomic splitting policy, especially as the range data inevitably consist of less-good specimens than the chosen reference material. Plants found only in Paleozoic strata have been omitted as being too remote when no connecting link at all in the Mesozoic has been postulated. No attempt is made below to describe plant fossils fully, for which see references to papers and to selected textbooks elsewhere: the purpose is to identify the record concerned clearly, and to draw attention to certain salient points of distribution and of character.

8.7–8.9   Permian and Early Triassic records

8.7   Glossopteris

The record of *Glossopteris* extends into Early Triassic rocks (Fig. 8.2), although this is at the end of the range of the very large group of *Gangamopteris* and *Glossopteris* leaves. Despite being first described from Australian specimens, much more material from South Africa and India has subsequently been presented with greater clarity.

Glossopteris 1. This is used here as a term for the large assemblages of Permian Sakmarian age material from the Lower and Middle Ecca successions of South Africa and equivalents elsewhere. From these beds Plumstead (1956, 1958) provided the first descriptions of many reproductive structures, *Scutum* and *Hirsutum*; she considered them to be bisexual, although this has subsequently been very clearly shown to be incorrect by, among others, Pant and Nautiyal (1984) describing *Ottokaria* (Fig. 8.3). All these extensive assemblages fall here into the Period Classification 260 (Permian Artinskian). Melville (1962, 1969) has claimed that some details of *Glossopteris* leaf venation are parallelled in several living angiosperm families. This is not disputed, but the general pattern of anastomosing venation (see Alvin and Chaloner 1970) is also closely similar to that of two Carboniferous pteridosperms (*Lonchopteris*, *Linopteris*) and at least two Mesozoic leaves (*Sagenopteris* and *Dictyozamites* in quite separate groups), and would be produced by a normal leaf-margin meristem. There is little reason to pursue Melville's claim further, unless some later link fossils are established; subsequent to *Glossopteris*, *Sagenopteris* leaves may suggest such a link but the rest of that plant does not support this.

Glossopteris 2. This is used for a different set of leaf species of late Permian age and is associated with new distinct reproductive structures (Fig. 8.4) particularly *Lidgettonia* Thomas (1958); this fossil is bisexual, although others such as *Plumsteadia* are ovule-bearing only. Some of the leaves continue into Early

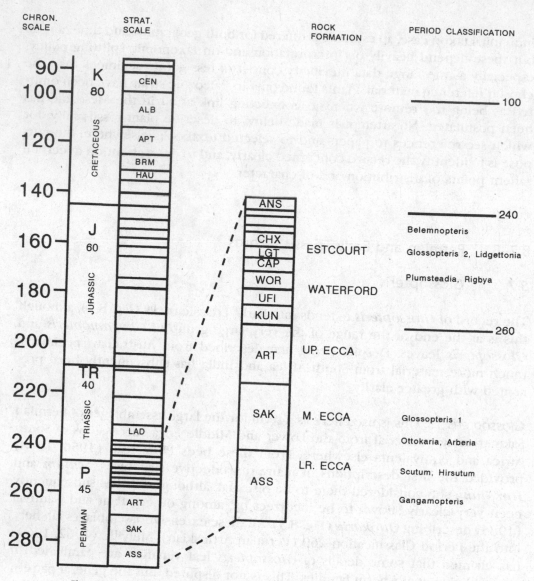

Fig. 8.2. The outline stratigraphic occurrence of *Glossopteris* and associated taxa in the Gondwana continents, particularly in South Africa. Occurrences are grouped in Period Classifications 260 and 240.

Triassic time and so naturally fall into the Period Classification 240 (Triassic Anisian). Like the earlier glossopterids, all these plants were apparently confined to the Gondwana region. An evolutionary pattern for leaf venations in the Glossopteridales has recently been proposed by Srivastava (1991), although he made no reference to Melville (1969). Retallack and Dilcher (1981a) made claims based on fertile *Dictyopteridium* particularly but admitted that considerable theoretical reduction and other changes are necessary to make a bitegmic ovule connection to angiosperms.

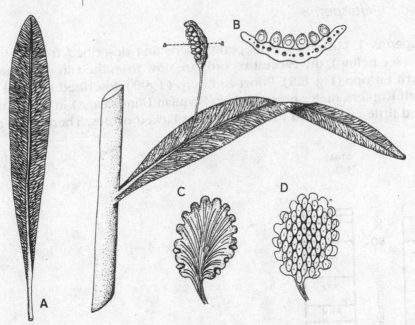

Fig. 8.3. (A) Early *Glossopteris* leaf showing venation and midrib, ×0.6 (from Stewart 1983). Reconstruction of *Glossopteris* leaf with *Ottokaria* attached fructification, ×0.6: (B) attachment and cross-section (a . . . .a) of *Ottokaria*; (C) abaxial (reference to stem) view; (D) adaxial (fertile) view (from Pant and Nautiyal 1984.)

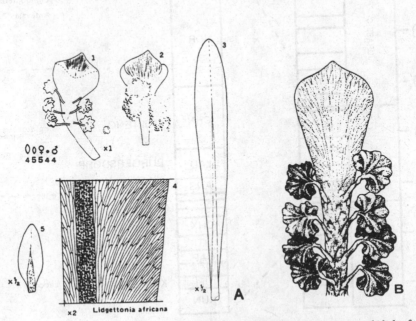

Fig. 8.4. Late Permian *Glossopteris* with fructification *Lidgettonia*: (A) leaf shape, midrib and venation (from Anderson and Anderson 1985); (B) reconstruction, ×1.5 (from Crane 1985).

8.8   *Peltasperms*

Peltasperms 1. Although peltasperms were first described from Late Triassic rocks (see below), the earliest records are now from the Late Permian rocks of western Europe (Fig. 8.5). Poort and Kerp (1990) described material from the difficult Kupferschiefer preservation (Permian Ufimian age) in which the leaves yielded little cuticle and the peltate discs lacked ovules. There is no reason to

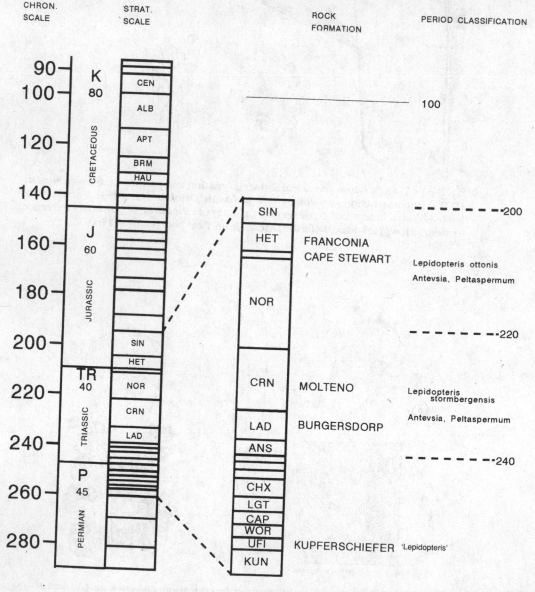

Fig. 8.5. Outline stratigraphic occurrence of the Peltasperm group with occurrences in Period Classifications 240, 220 and 200.

question their identification, although these authors chose the unhelpful nomenclatural policy of erecting 'natural genera'; this tends to obscure the nature of individual records, which are always based on separately dispersed organs, consequently their new combinations will be better dissociated from data and used only in interpretations. These descriptions support the contention of Meyen (1987) that other plant groups such as cardiolepids and corystosperms should be placed with peltasperms into an Order Peltaspermales, linking Paleozoic and Mesozoic floras; the individual groups are treated separately here to give them adequate attention and to avoid submerging their interesting distributions in space and time. The southern hemisphere occurrences of peltasperms are from the Triassic Ladinian Burgersdorp Formation and from the Triassic Carnian Molteno Formation of South Africa. Anderson and Anderson (1983, 1989) figure the bipinnate frond *Lepidopteris stormbergensis*, together with *Antevsia* (a pollen-producing organ) and *Peltaspermum* with ovules and seeds (Fig. 8.6). These plants were by no means dominant in the Molteno flora (Fig. 8.7); the seeds produced were small and the stems slender, probably indicating wide dispersal potential (Anderson and Anderson 1989, p. 44).

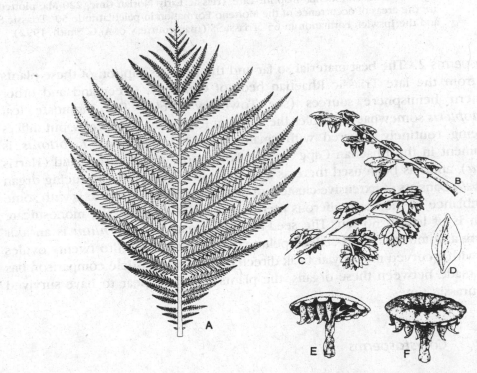

Fig. 8.6. The peltasperms: (A) *Lepidopteris stormbergensis*, ×1/4, from the Triassic Molteno Beds of South Africa (from Anderson and Anderson 1989); (B) *Autunia* megasporophyll, ×1.5; (C) *Antevsia zeilleri*, microsporophyll, ×1; (D) Monosulcate pollen of *Antevsia*, ×350; (E,F) Peltasperm peltate organ bearing ovules, ×1.5 (from Crane 1985).

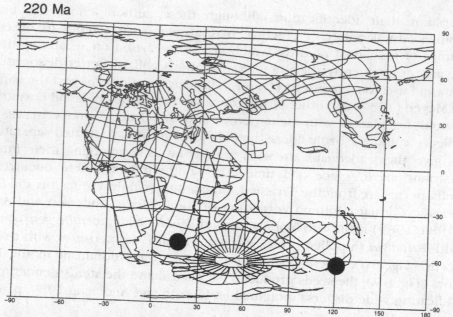

Fig. 8.7. Paleocontinental map for Late Triassic Early Norian time, 220 Ma; plotted are the areas of occurrence of the Molteno Formation in paleolatitude 50° Triassic S and the Ipswich Formation in 65° Triassic S (map courtesy of A. G. Smith, 1992).

Peltasperms 2. The best material so far and the first description of these plants was from the late Triassic Rhaetian beds of Germany, Greenland and other northern hemisphere sources (see Townrow 1960). The bipinnate leaf *Lepidopteris* somewhat resembles the Permian leaf-genus *Callipteris*, but differs in being routinely covered with small blister-like swellings; *L. ottonis* is prominent in the Rhaetian Cape Stewart Formation of East Greenland (Harris 1932a), and has been used there as a zone fossil. The pollen-producing organ *Antevsia* consists of extensive clusters of 'sporangia' on a slender axis with some resemblance to the Carboniferous pteridosperm *Crossotheca*; the monosulcate pollen is of length 36 µm. The seed-bearing organ *Peltaspermum* is an axis bearing a number of short-stalked peltate discs; a disc has up to twenty ovules each with a curved micropylar beak directed outwards. Cuticle comparison has been made between these organs; the plants do not appear to have survived into Jurassic times.

8.9   Corystosperms

The leaves and cuticles of this group of common southern hemisphere plants (Fig. 8.8) have been very well studied by Anderson and Anderson (1989). Much less is known of its reproductive organs, first described by Thomas (1933) from Natal as *Umkomasia* with recurved cupules on a branching axis, and *Pteruchus* with branched 'sporangia' again somewhat resembling the Carboniferous pteri-

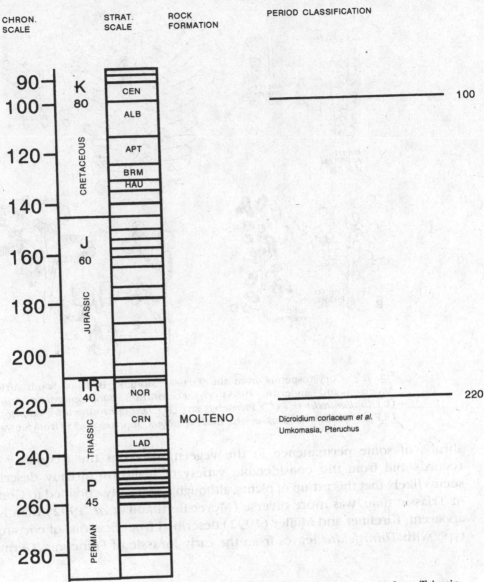

Fig. 8.8. The main age of occurrence of the Corystosperm group in Late Triassic time (Period Classification 220).

dosperm *Crossotheca* (Fig. 8.9); Thomas (1933) did not illustrate the small saccate pollen which was figured by Townrow (1962*b*) like that of *Caytonia* but showed a more pronounced sulcus and, according to Couper (1958), of about twice the size (this size may have been exaggerated by Couper's preparation methods). Other material was subsequently very fully described by Townrow (1962*a*) from Tasmania.

It has been suggested by Archangelsky (1968*a*) that the rather common Gondwana Triassic wood *Rhexoxylon* (known to up 25 cm in diameter) was an organ of the corystosperms, and that consequently the plants were woody

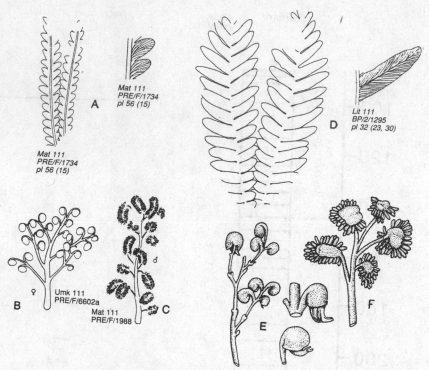

Fig. 8.9. Corystosperms from the Triassic Molteno Beds of South Africa (from Anderson and Anderson 1983): (A) *Dicroidium odontopteroides orbiculoides*; (B) *Umkomasia* sp.; (C) *Pteruchus* sp.; (D) *D. odontopteroides odontopteroides*; (E,F) restorations of *Umkomasia and Pteruchus*, approx. ×0.75 (from Stewart 1983).

shrubs of some permanence in the vegetation. From the frequency of leaf records and from the considerable variety of leaf-form already described, it seems likely that this group of plants, although apparently confined to Gondwana in Triassic time, was more diverse (Meyer-Berthaud *et al.* 1992) than is so far apparent; Kirchner and Müller (1992) described fructifications of corystosperm type with *Thinnfeldia* leaves from the early Jurassic of Franconia, Germany.

8.10–8.16 Late Triassic records

8.10   Czekanowskia (Leptostrobus) 1

Although the principal information on this group of fossils comes from Jurassic and Cretaceous rocks, there is a presence of leaves in the Triassic Rhaetian and Jurassic Hettangian of the Cape Stewart Formation of East Greenland. *Czekanowskia* is a short shoot (see Section 8.23 and Figs. 8.33, 8.34) with very narrow leaves, each entered by one vein with subsequent dichotomies of both the leaf lamina and vein; *Hartzia* is a simpler leaf with one dichotomy; *Phoenicopsis* consists of blunt linear leaflets clustered also in a short shoot. The fertile structures *Staphidiophora* and *Leptostrobus longus* are associated with the leaves, but not in a particularly favourable preservation. Meyen (1987) regarded these plant fossils as descended from peltasperms.

8.11   *Triassic Carnian* Sanmiguelia lewisii

This extraordinary plant fossil of which the actual dating (Fig. 8.10) has been improved to Earliest Norian by Litwin *et al.* (1991) was originally described by

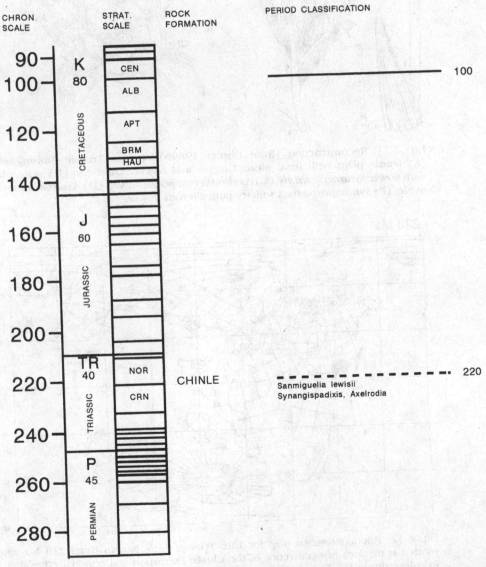

Fig. 8.10. The age of occurrence of the *Sanmiguelia* group of plant fossils in south-western North America (Period Classification 220).

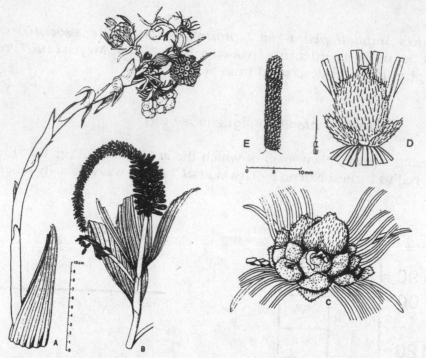

Fig. 8.11. Reconstruction (from Cornet 1989*b*) of Late Triassic *Sanmiguelia*: (A) female plant with basal pleated leaves and flower *Axelrodia*; (B) male plant with flower *Synangispadixis*; (C) *Axelrodia* composite flower; (D) *Axelrodia* solitary ovule; (E) *Synangiospadixis* solitary polleniferous flower.

220 Ma

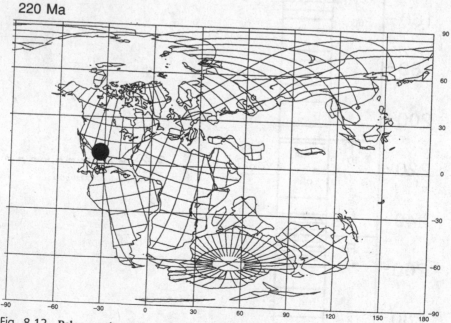

Fig. 8.12. Paleocontinental map for Late Triassic Early Norian time, 220 Ma ago; plotted is the area of occurrence of the Chinle Formation and regional equivalents in paleolatitude 10° Triassic N (map courtesy of A. G. Smith, 1992).

Brown (1956) from impressions of large pleated leaves found in south-western Colorado and believed to be palm-like; more information was added from new Texan material by Ash (1976) and Tidwell *et al.* (1977). Cornet (1986, 1989*b*) described much more complete and fertile material from Sunday Canyon in north-west Texas; robust stems were described in a position of growth and also fallen, although the attached leaves were all damaged. *Axelrodia burgeri* is a name given to an ovuliferous inflorescence found in close association with these stems (Fig. 8.11); the description was of flowers of two kinds showing closed carpels with pairs of basal ovules and apical stigmas. The male inflorescence *Synangispadixis tidwellii* was described as a spike bearing hundreds of spirally inserted flowers without perianth; each flower was described as a pair of anthers containing simple monosulcate pollen (Fig. 8.11). The very enthusiastic description (Cornet 1989*b*) referred to transmission tissue in the carpel and to development of two cotyledons; it was not possible to establish whether the plant was monoecious. This plant fossil, or this incomplete set of organ fossils, is sufficiently unusual to have frustrated serious attempts at comparison with other plants, fossil or living. It has been vigorously studied but clearly even more good specimens are needed. Overall, *Sanmiguelia lewisii* signals a warning (Fig. 8.12) that knowledge of paleoequatorial Mesozoic Pollenifera is inadequate for the interpretation expected from it; many more such surprises should be anticipated and sought, and classification should be applied only very lightly at this early stage.

8.12   Dinophyton

Krassilov and Ash (1988) described this curious small plant from the Chinle Formation and Dockum Paper Coals (Triassic Carnian) of Arizona somewhat optimistically as protognetalean, although the interest lies in the fossil rather than in any attribution. The description is of large numbers of small 'pinwheel' samaras, each with an ovule in a cup, associated with *Brachyphyllum* shoots, and with *Alisporites* pollen similar to that of the corystosperms. This is as yet an isolated occurrence, from a rock formation that has already contributed surprises; it is an example of a group of Mesozoic Pollenifera that will probably become regarded as more diverse when individual discoveries are released from the traditional unhelpful 'conifer' attribution.

8.13   Caytonia (Sagenopteris) *1*

Fossils from the Thaumatopteris Zone (Jurassic Hettangian; about 205 Ma) of the Cape Stewart Formation, Scoresby Sound, East Greenland, were described by Harris (1932*b*); these included three species of *Sagenopteris* leaves, the new cupule *Caytonia thomasi* (Harris 1933), a male organ *Antholithus* well

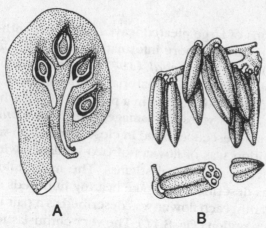

Fig. 8.13. Reconstructions of Early Jurassic Greenland plants, approx. ×8: (A) *Caytonia thomasi*, famous diagram originally from Harris (1932*b*, 1933) but of composite origin; (B) *Caytonanthus kochi* (from Stewart 1983).

preserved but detached (Fig. 8.13), scattered very small saccate pollen, and some isolated *Amphorispermum* seeds. These were described very accurately and skilfully, but with diagrams made in the light of the information from the better preserved material of Thomas (1925) from Yorkshire rocks dated 30 million years later. The details of this earliest occurrence of *Caytonia* and *Sagenopteris* provide a potentially important evolutionary link from one of the earlier Triassic plant groups through to the remainder of the Mesozoic era but this needs redevelopment. For distribution, see Fig. 8.23 (below).

8.14   Hirmeriella: 'Classopollis' plants 1

The genus *Hirmeriella* Hörhammer of the family Cheirolepidiaceae with type species *H. muensteri*, of late Triassic Rhaetian age from Germany but known

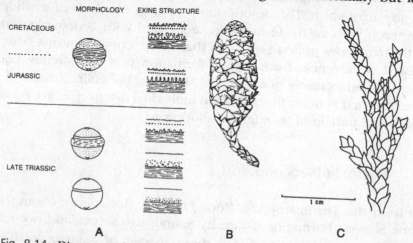

Fig. 8.14. Diagrams relating to the Late Triassic '*Classopollis*' plant: (A) dispersed pollen *Classopollis*; (B) ovuliferous cone *Hirmeriella muensteri* (from Jung 1968); (C) shoot of *Hirmeriella muensteri* (A and C from Alvin 1982).

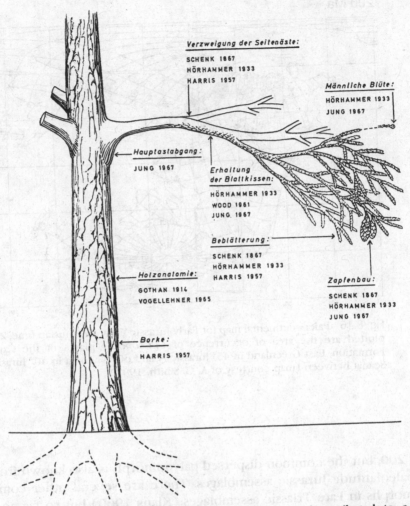

Verzweigung der Seitenäste:
SCHENK 1867
HÖRHAMMER 1933
HARRIS 1957

Männliche Blüte:
HÖRHAMMER 1933
JUNG 1967

Hauptastabgang:
JUNG 1967

Erhaltung der Blattkissen:
HÖRHAMMER 1933
WOOD 1961
JUNG 1967

Beblätterung:
SCHENK 1867
HÖRHAMMER 1933
HARRIS 1957

Holzanatomie:
GOTHAN 1914
VOGELLEHNER 1965

Zapfenbau:
SCHENK 1867
HÖRHAMMER 1933
JUNG 1967

Borke:
HARRIS 1957

Fig. 8.15. The sources from which information has contributed to a general reconstruction of the early '*Classopollis*' plant (Triassic–Jurassic). From below upwards on the trunk (and left to right on the branch): bark, wood of trunk, major branching, division of branches, leaf scars, leafy twigs, female cones, male cones (?connection). (From Jung 1968 and see references therein.)

(Fig. 8.14) in western Europe at least from Poland to Wales (Harris 1957), is described as an arborescent plant with small helically arranged leaves up to 3 mm long (inevitably referred to as a 'conifer'). Both a female cone with ovuliferous scales each with a pair of inverted ovules, reinterpreted by Jung (1968) and by Clement-Westerhof and Van Konijnenburg-van Cittert (1991), and a male cone with peltate-headed microsporophylls each with two pollen sacs, have been found attached to the plant (Fig. 8.15). The *in situ* pollen is 20–29 µm in diameter and of *Classopollis* type; nomenclature code purists (see Traverse 1988) prefer the name *Corollina* on priority, but as so often in such cases this provides no information. This plant record falls in the Period Classi-

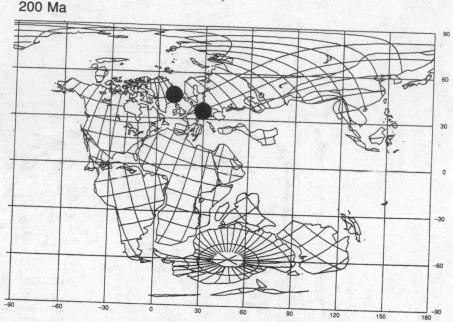

200 Ma

Fig. 8.16. Paleocontinental map for Early Jurassic Early Sinemurian time, 200 Ma ago; plotted are the area of occurrence of *Classopollis* plants of the Cape Stewart Formation, East Greenland in 45° Jurassic N and of Franconia in 40° Jurassic N, with Scania between (map courtesy of A. G. Smith, 1992).

fication 200, but the common dispersed palynomorph is also known from most lower-paleolatitude Jurassic assemblages. There are several other comparable palynomorphs in Late Triassic assemblages (Klaus 1960) but so far no further mega-fossils (Fig. 8.16). The occurrence of this dispersed pollen in tetrads has been reported on numerous occasions (Pocock *et al.* 1990), but it is not known how the frequency of this relates to the severity of oxidation treatment during palynomorph recovery; frequently before about 1970 and certainly by the late fifties (Couper 1958) the treatment then in use is now known to have been unnecessarily severe, but was by no means always adequately described and thus accounted for. Pollen-bearing organs of this kind have been found associated with *Brachyphyllum scottii* from the Scottish Lower Lias (Kendall 1949), with *Cupressinocladus ramonensis* from the Israeli Lower Lias (Lorch 1968), and with *Pagiophyllum araucarianum* from the Upper Lias of England (Harris 1979) with female *Hirmeriella estonensis* attached, and also from the Upper Lias of France (Barale 1981).

8.15   Irania hermaphroditica *Schweitzer 1977*

This remarkable and well-described fossil from the Triassic Rhaetian of Iran (Fig. 8.17) has not yet been recorded again elsewhere. The male and female structures borne on different parts of the same axis make a relatively small flower of about 3–4 cm length, and there is no perianth (Fig. 8.18). The leaf *Desmiophyllum* is associated but not connected, and there is apparently no

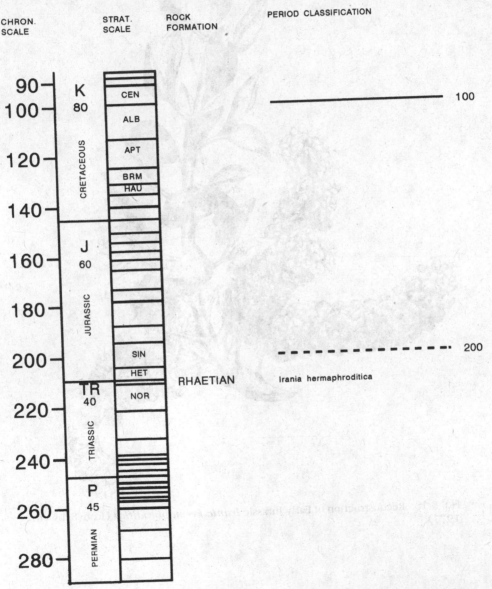

Fig. 8.17. The age of occurrence of the isolated Triassic plant *Irania* from southern Asia (Period Classification 200).

Fig. 8.18. Reconstruction of Early Jurassic *Irania hermaphroditica* (from Schweitzer 1977).

cuticle available. The existing evidence does not make this plant fossil a good angiosperm candidate and it was not so claimed by the author, but it adds most interestingly to the early Jurassic diversity of the Pollenifera; it provides a further indication of how unusual Mesozoic floras can be when explored without bias and when freed from any necessity to explain them in terms of living gymnosperms. Meyen (1987) places *Irania* close to *Czekanowskia* and *Leptostrobus*, but such classificatory grouping does not assist with interpretation.

8.16 Williamsonia *group 1*

The *Williamsonia* group consists of plant fossils that bore substantial flowers connected to or associated with robust pinnate leaves with paracytic stomata (often referred to as syndetocheile stomata), and slender stems. Since Thomas and Bancroft (1913) distinguished their cuticle the group has been recognised as distinct, generally under the name of the 'Order Bennettitales' (or sometimes confusingly as the 'Order Cycadeoidales' as described by Delevoryas and Gould (1973)), but as there is no need for such a hierarchical arrangement it may now be preferable to avoid both names as neither of them is required in Period Classifications. The genus *Cycadeoidea* is valid and is considered separately (Section 8.30) below.

The earliest fossils include *Wielandiella angustifolia* (originally *Wielandia*) from the late Triassic Rhaetian of southern Sweden; it was a small ovulate flower (<2 cm in diameter) firmly associated with the pinnate leaf *Anomozamites minor* with the paracytic stomata of the group. The frequently reproduced restoration of this plant is unintentionally misleading as no flower has yet been observed in attachment; the restoration persists because nobody wishes to destroy an attractive portrait.

Rocks of similar age in East Greenland, in the Lepidopteris Zone of the Cape Stewart Formation (Triassic Rhaetian), indicate considerable possible variety of such plants in the presence of four species of *Anomozamites* and nine of *Pterophyllum*. The reproductive structures, however, are fragmentary, under such names as *Vardekloeftia* and *Hydropterangium*, which may ultimately be paired with leaves, although *Cycadocephalus*, which has been even longer in the literature, has now been placed (Chaloner *et al.* 1991) with the ferns. Excellent new Triassic Rhaetian material of *Vardekloeftia* has now been presented by Raunsgaard Pedersen *et al.* (1989) with details of seeds, attachment scar, interseminal scales and cuticles but without establishing attachment to other organs (Fig. 8.19).

Elsewhere, in the Late Triassic rocks of Lunz in Lower Austria, several other fragmentary organs have been recorded under the names *Haitingeria* (microsporophyll), *Bennetticarpus* (ovulate), *Cycadolepis* (scales) and *Westerheimia* (axes), with numerous leaf species. The genus *Sturiella* was erected by Kräusel (1949) for a flat sunflower-like reproductive structure from Lunz.

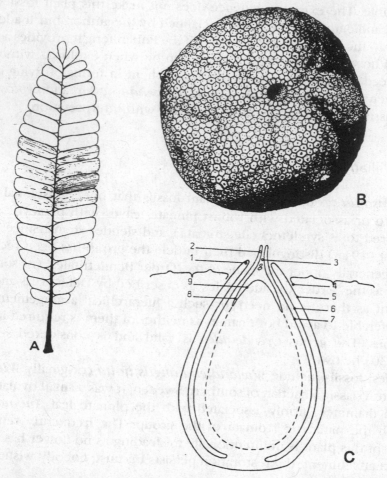

Fig. 8.19. The Late Triassic–Early Jurassic members of the Williamsonia group (Bennettitales): (A) *Pterophyllum kochii*, ×0.3 (from Crane 1985); (B) reconstruction of *Vardekloeftia*, showing protruding apices of large seeds and attachment scar, ×2; (C) longitudinal section of one mature seed (1, cupule; 2, integument; 3, micropyle; 4, interseminal scale; 9, nucellus) (B and C from Pedersen *et al.* 1989).

The group is thus very diverse in its early occurrences and probably continued collecting is still available in the Triassic rocks concerned.

8.17–8.24 Jurassic Records

8.17   Eucommiidites plant 1

The original description of this dispersed pollen *E. troedssonii* was by Erdtman (1948) from Jurassic Hettangian rocks of southern Sweden, but Scheuring

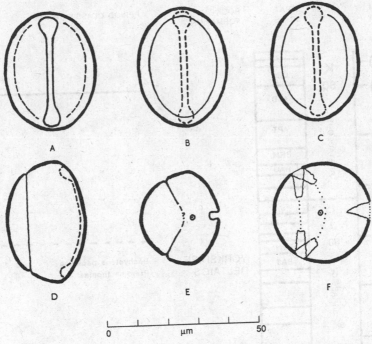

Fig. 8.20. The dispersed Early Jurassic pollen *Eucommiidites troedssonii*, ×1000: (A) presumed distal view; (B,C) proximal views with complete, or broken ring sulcus; (D) lateral view; (E,F) end view unexpanded, and expanded (main distal furrow to right) (from Hughes 1961).

(1970) extended the range back into the Late Triassic of the Swiss Jura Mountains. Reymanovna (1973) observed these pollen grains in the long micropyle of an isolated ovule from Jurassic Aalenian rocks of Grojec, Poland, paralleling earlier similar observations from Early Cretaceous rocks (see Section 8.32). The observations of Couper (1958) and Hughes (1961, 1976), which established that the grain was not tricolpate (and thus not angiospermous) but distally monosulcate and proximally zonosulcate (usually interrupted), were made on Jurassic Bathonian material (Fig. 8.20). In fact there is as yet no other pre-Cretaceous megafossil or mesofossil information, although as with several other innovations the palynologic beginnings came in Late Triassic time.

8.18   Pachypteris

A simply pinnate frond *Pachypteris papillosa* with a very thick cuticle is recorded from the Yorkshire Deltaic rocks of Jurassic Bajocian age (Fig. 8.21). Associated with it (Harris 1964) is a male organ *Pteroma thomasi* with bisaccate pollen closely resembling *Alisporites* (*Pteruchipollenites*) *thomasi* (Couper 1958); no female organ is yet known. Another accompanying leaf species

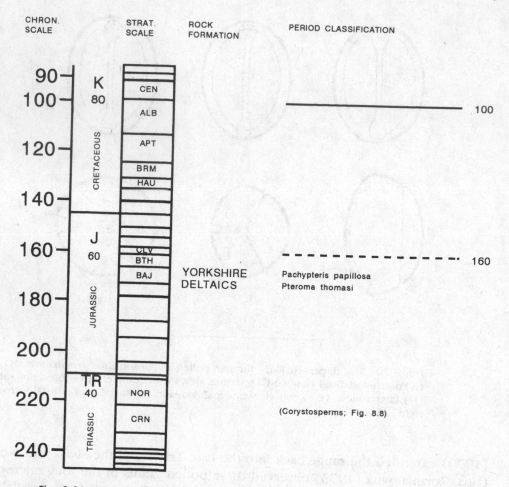

Fig. 8.21. The Jurassic age of occurrence of the principal well-described records of *Pachypteris* from England (Period Classification 160). Compare Fig. 8.8 for the corystosperms.

P. lanceolata is bipinnate. All these plant fossils are relatively similar to the corystosperms, with *Dicroidium* (leaf) and *Pteruchus* (male organ) from the Molteno Formation of Triassic Carnian age from South Africa (see Section 8.9), and could be candidates for evolutionary successors of the corystosperms; such a connection is suggested by Kirchner and Müller (1992). In Yorkshire *Pachypteris* is always found in strata containing dinoflagellates (Van Konijnenburg-van Cittert 1971) and is therefore to be considered to be of marine margin habit, which is also compatible with the presence of a very thick cuticle (Fig. 8.22). There are other Jurassic records from Central and Eastern Europe, Caucasian Georgia and Kazakhstan, and most of these are believed to have been from littoral deposits (Vakhrameev 1991, p. 249). The earliest *Pachypteris* record may be from the Jurassic Toarcian deposits of Trans-Caspian Mangyschlak, but has the same littoral association.

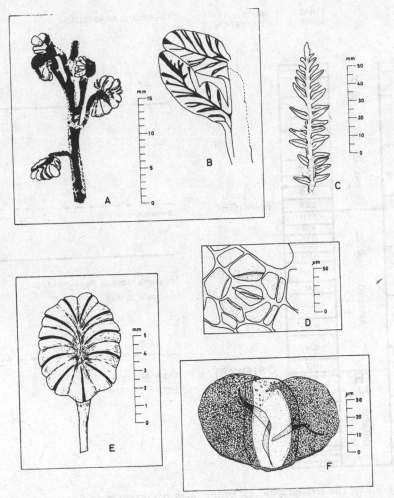

Fig. 8.22. Jurassic *Pachypteris*: (A,E) *Pteroma thomasi*; (A) part of fertile specimen; (E) restoration of single head; (B–D) *Pachypteris papillosa*; (D) stoma from lower cuticle; (F) *P. thomasi* pollen (from Hughes 1976).

8.19   Caytonia (Sagenopteris) 2

This group of fossils (see also Section 8.13) illustrates well the problem that virtually all the critical descriptive information (Fig. 8.23) comes from many specimens of the one excellent chance type of preservation at the original locality in East Yorkshire (Cayton Bay Formation; Early Bajocian; *c.* 169 Ma) as described by Hamshaw Thomas (1925). Thomas figured the cupule as angiospermous with a receptive stigmatic surface, although this interpretation was withdrawn when the individual ovules within were shown to have separate

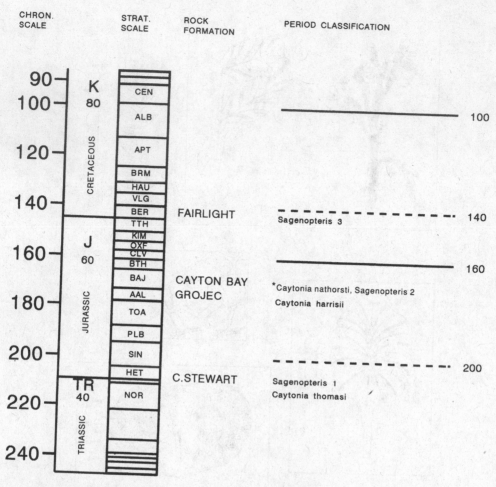

Fig. 8.23. The stratigraphic age of the main occurrences of *Caytonia/Sagenopteris*, of which *Caytonia* 2 from the Jurassic Bajocian of Yorkshire provides most of the detailed morphologic knowledge, from Period Classification 160.

micropyles; the whole structure was further figured and explained by Harris (1940, 1941). The fossils were a palmate leaf with anastomosing venation *Sagenopteris phillipsi*, *Caytonia nathorsti* (a megasporophyll with numerous cupules), *Caytonanthus arberi* (a male inflorescence with pollen), the seed *Amphorispermum pullum* and abundant dispersed very small (sulcus length 15 μm) saccate pollen *Vitreisporites*; the unusual mummified compression that provides excellent preservation of cuticles (Fig. 8.24). The fossils are common (rather than abundant) in well-exposed beds and so have been collected also by other observers (Fig. 8.25). Thomas (1925) came close to solving an angiosperm origin problem with this discovery of a virtually closed cupule, and may

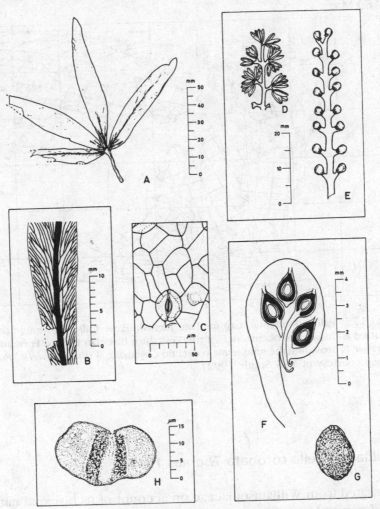

Fig. 8.24. Jurassic Bajocian *Caytonia* from Yorkshire: (A–C) *Sagenopteris phillipsi* leaf, venation and cuticle; (D) *Caytonanthus arberi*; (E,F) *Caytonia nathorsti* megasporophyll, and restoration in longitudinal section of fruit; (G) *Amphorispermum* seed; (H) *Caytonanthus* pollen (from Hughes 1976).

yet be shown to have been nearer than even he supposed in his later life-time.

The *Caytonia harrisii* fossils described by Reymanovna (1973) from Grojec near Krakow, Poland, were of Jurassic Aalenian age, slightly older than the Yorkshire fossils. The several species of *Sagenopteris* leaves from the Veneto flora of northern Italy (Wesley 1966) were of Late Pliensbachian age, older again but without reproductive structures. Recently material has been discovered in Antarctica (Rees 1993).

160 Ma

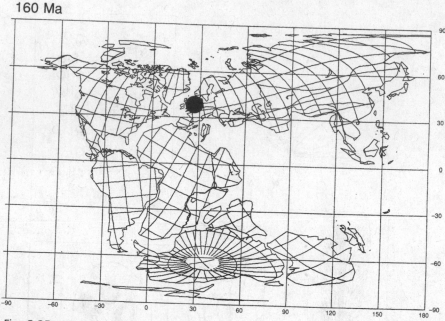

Fig. 8.25. Paleocontinental map for Late Jurassic Early Callovian time, 160 Ma ago; plotted is the area of occurrence of the Yorkshire Bajocian Deltaic Formation in 40° Jurassic N providing the unique main data on *Caytonia, Williamsoniella, Pachypteris* (map courtesy of A. G. Smith, 1992).

8.20   Williamsoniella coronata *Thomas 1915*

This was separated from Williamsoniaceae on account of its bisexual nature, its leaf genus *Nilssoniopteris vittata* with paracytic stomata and its small size of flower (<2.5 cm in diameter). The specimens come from the Cayton Bay Formation of East Yorkshire of Jurassic–Bajocian age (Fig. 8.26) and were reviewed by Harris (1944). Another species from the earlier Bajocian outcrops of Whitby was named *W. papillosa* by Cridland (1957). Harris (1974) reinvestigating *W. lignieri* Nathorst 1909, a specimen stored in Stockholm and originally described under *Williamsonia*, decided that these two species of *Williamsoniella* should be united under *W. lignieri* with a leaf species *N. major*, and perhaps more significantly with the pollen species *Exesipollenites scabratus* with a diameter of 25 μm and a single pore as the only feature on an otherwise spherical pollen grain. Van Konijnenburg-van Cittert (1971) described the *in situ* pollen of *W. coronata* from Cayton Bay as monocolpate, but Harris (1974, p. 132) indicated that he regarded this as an *Exesipollenites* shape with similar small diameter, and that the frequent compressional exine folds merely gave the appearance of a sulcus. A single small flower *W. valdensis* was recorded from Cretaceous Berriasian rocks of Sussex

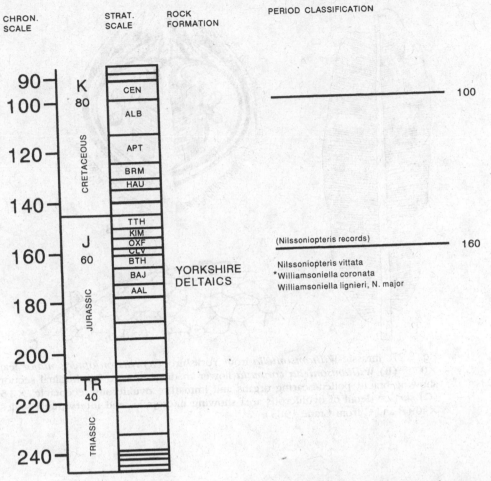

Fig. 8.26. The stratigraphic age of the excellent but very restricted source of information on *Williamsoniella* in Period Classification 160, Jurassic Bajocian.

(Edwards 1921), but is now relegated to *Bennettistemon* by Watson and Sincock (1992), and should not be attributed to this small group.

Although species of *Williamsoniella* are occasionally mentioned elsewhere, e.g. *W. karataviensis* and *W. czochaiensis* by Vakhrameev (1991, p. 69), these really refer to records of the taeniopterid-type leaves *Nilssoniopteris*. The diagnostic characters of *Williamsoniella* (Fig. 8.27) have been recorded only from Yorkshire Jurassic Bajocian rocks, and as these characters are distinct from those of all other contemporaneous plants this group should be regarded as entirely separate from the plant of *Williamsonia/Weltrichia*. It is only unfortunate that Thomas (1915) did not propose an entirely independent name that would have made this clearer, although at that time the views of Nathorst, Seward and Wieland would probably have prevailed against this. If there is any connection

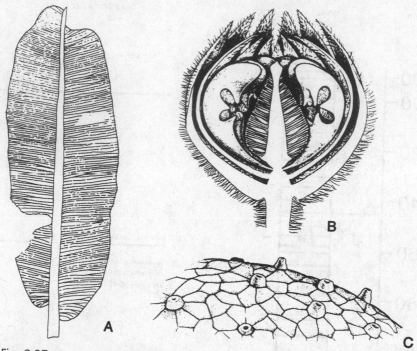

Fig. 8.27. Jurassic *Williamsoniella* from Yorkshire: (A) *Nilssoniopteris major* leaf, ×0.75; (B) *Williamsoniella coronata* flower in diagrammatic longitudinal section showing bracts, pollen-bearing organs and immature ovuliferous receptacle, ×2.5; (C) surface detail of ovuliferous area showing micropyles and interseminal scales, ×40 (B and C from Crane 1985).

with later plant fossils, the comparison might be with *Cycadeoidea*, which is also bisexual (see e.g. Crane 1985).

8.21   Williamsonia *Group 2*

For continuity see also Section 8.16. Numerous relatively well-preserved large flowers (up to 5 cm or more in diameter) have been described from Jurassic Bajocian rocks of East Yorkshire (Fig. 8.28). Knowledge of them has been summarised by Harris (1969), who restricted the name *Williamsonia* to unisexual ovuliferous fossils; he employed *Weltrichia* (rather less common fossils but of similar or slightly larger size) for male flowers only. Several species of both flowers are principally preserved as compressions in fine sediment, and are frequently found with good cuticle that has been matched with that of species of the associated pinnate leaf genera *Pterophyllum*, *Ptilophyllum*, *Oto-*

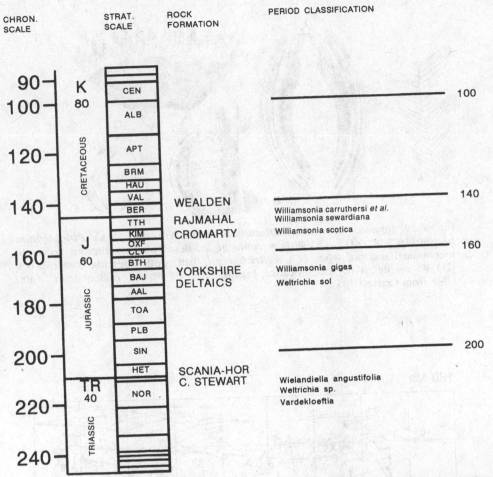

Fig. 8.28. The stratigraphic range of information from fossils of the Williamsonia group as at present envisaged. Period Classifications from 200 to 140 inclusive.

zamites and *Zamites* (Fig. 8.29). The pollen taken from *Weltrichia sol* of Jurassic Bajocian age was thin-walled monosulcate of length 46 μm which appears to be representative of the *Williamsonia* group; this is clearly distinct from the monoporate pollen (25 μm in diameter) of *Williamsoniella* (see Section 8.20).

Watson and Sincock (1992) describe seven species of *Williamsonia* (including *Bennetticarpus*) from Cretaceous Berriasian rocks of the Lower Wealden Beds near Hastings, Sussex, and provide some interesting restorations of them; although several provided cuticle, the preservation was not as complete as for the Yorkshire Jurassic specimens whence the main information on the group is derived (Fig. 8.30 shows other localities of the Euro-Sinian province described by Vakhrameev (1991)).

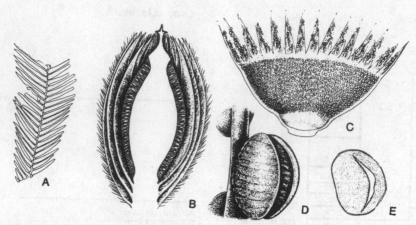

Fig. 8.29. Jurassic Bajocian *Williamsonia* from Yorkshire. (A) *Ptilophyllum pectinoides* leaf, ×0.3; (B) *Williamsonia gigas*, developing ovuliferous flower in longitudinal section, ×0.5; (C) *Weltrichia sol*, 'half' pollen-bearing flower, ×0.3; (D) *W. sol* detail of pollen-bearing organs, ×6; (E) *W. sol.* pollen grain, ×400 (B–E from Crane 1985).

Fig. 8.30. Paleocontinental map for Late Jurassic Early Callovian time, 160 Ma ago; Euro-Sinian Province described by Vakhrameev (1991), conjectural in North America: 1, England; 2, France, Iberia; 3, Germany; 4, South Russia; 5, West Siberia; 6, Middle Asia; 7, West China; 8, Japan; areas of *Williamsonia* foliage occurrence (map courtesy of A. G. Smith, 1992).

8.22   Hirmeriella: 'Classopollis' plants 2

See also Section 8.14. From the famous Yorkshire Deltaic succession of Jurassic Bajocian–Bathonian age (Fig. 8.31), there are *Brachyphyllum crucis* with male and female organs cuticle identified by Kendall (1947), and *Pagiophyllum maculosum* (see Harris 1979) with *Hirmeriella kendalliae* described by Van Konijnenburg-van Cittert (1987); both have relatively thin cuticles (about 6 μm) and *in situ* pollen of '*Classopollis*' type. The term 'conifer', which has for many years been applied to these and various other plants with generally similar leaves (Fig. 8.32), should be consciously avoided for '*Classopollis*' plants, particularly in order to separate them more clearly from plants bearing *Araucariacites* and other putatively 'conifer' pollen. Vakhrameev (1991 and earlier)

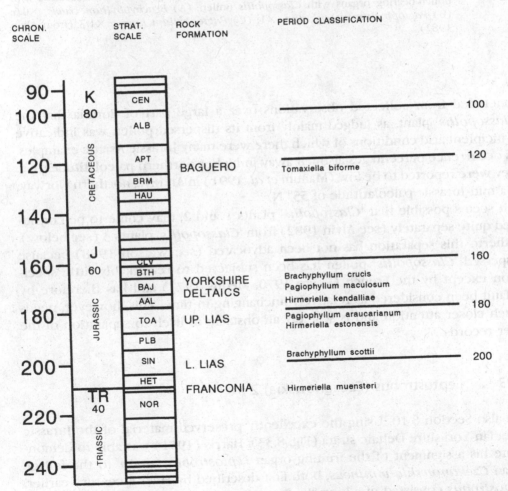

Fig. 8.31. The stratigraphic age of the principal occurrences of the *Hirmeriella/Classopollis* groups 1 and 2 (Period Classifications 200 to 160), together with a geographically isolated record from Period Classification 120.

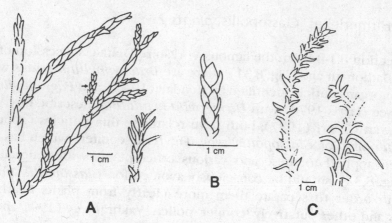

Fig. 8.32. Shoots of Jurassic Bajocian plants of Yorkshire that are associated with pollen-bearing organs with *Classopollis* pollen: (A) *Brachyphyllum crucis*, ×0.6; (B) *Pagiophyllum maculosum*, ×1; (C) *Pagiophyllum kurrii*, ×1/3 (from Alvin 1982).

concluded from collected observations over a large part of Eurasia that the '*Classopollis*' plant, as judged mainly from its dispersed pollen was indicative of incipient arid conditions of which there were many Jurassic Asiatic examples. Its occurrence percentages also fell away in higher northern paleolatitudes, e.g. they were reported to be low (Manum *et al.* 1991) in Andøya, northern Norway, at a mid-Jurassic paleolatitude of 55° N.

It seems possible that '*Classopollis*' plants 1 and 2, may come to be considered quite separately (see Alvin 1982) from '*Classopollis*' plants 3 (see below). Hitherto this separation has not been advocated (see Watson 1988) because dispersed '*Classopollis*' pollen has been subjected to remarkably little speciation except by the French (Reyre 1970, Medus 1977) and has therefore by default been considered as relatively unchanging in time. This, however, needs much closer attention and is scarcely an obstacle in itself to separation of the later records.

8.23   Leptostrobus (Czekanowskia) 2

See also Section 8.10. Using the excellently preserved material of the Jurassic Bajocian Yorkshire Deltaic strata (Fig. 8.33), Harris (1951) was able to demonstrate his assignment of the fruiting organ *Leptostrobus cancer* to the foliage organ *Czekanowskia vimineus*, both first described by Heer a century earlier. *Leptostrobus* consisted of a long slender axis with well-spaced bivalved lateral capsules, each valve with a row of three or four seeds, thus indicating an inde-

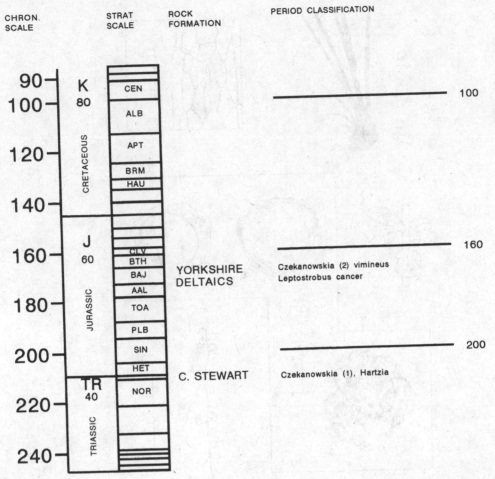

CHRON. SCALE   STRAT SCALE   ROCK FORMATION   PERIOD CLASSIFICATION

YORKSHIRE DELTAICS

Czekanowskia (2) vimineus
Leptostrobus cancer

C. STEWART   Czekanowskia (1), Hartzia

Fig. 8.33. The stratigraphic age of the best described occurrences of *Leptostrobus/ Czekanowskia* 1 and 2 from Period Classifications 200 and 160.

pendent group, (Harris *et al.* 1974) clearly separate from the traditional classification of Mesozoic 'ginkgo' leaves (Fig. 8.34). It so happens that the Yorkshire occurrence lies at the extreme south (35° Jurassic N) and west edge of a large Eurasian Jurassic distribution in both the Siberian and parts of the Euro-Sinian provinces described by Vakhrameev (1991). These fossils and the leaf *Phoenicopsis* occur abundantly in the South Urals, Trans-Caspian, Fergana, Mongolia, Lena Basin and Amur Basin regions. Also found in these areas is the presumed pollen-bearing organ *Ixostrobus*; the simple monosulcate pollen has not been characterised separately from other similar pollen, using the light microscope.

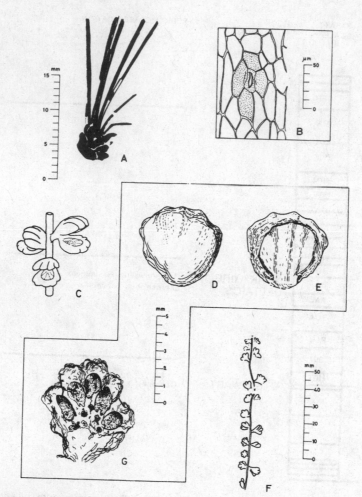

Fig. 8.34. Jurassic Bajocian Czekanowskiales from Yorkshire: (A,B) '*Solenites*' *vimineus*; (A) shoot; (B) cuticle; (C–E) *Leptostrobus* fructifications; C, *L. cancer*, reconstruction; D,E, outside and inside of one valve. Early Jurassic Czekanowskiales from East Greenland; (F) *Leptostrobus longus*; (G) capsule enlarged (from Hughes 1976).

8.24   *The* Pentoxylon *plant*

These plant fossils were first described from the Rajmahal Hills (Fig. 8.35) in eastern India as of early to mid-Jurassic age, and have been beneficially regarded as unclassifiable (Bose *et al.* 1985) by traditional botanical criteria. The leaves *Nipaniophyllum* were small (7×1 cm) and of *Taeniopteris* type; the slender axis (1 cm in diameter), found permineralised (*Pentoxylon*), contained five prominent mesarch bundles of pycnoxylic secondary wood. *Carnoconites*, the ovule-bearing fruit-like organ resembled a small mulberry, although a large

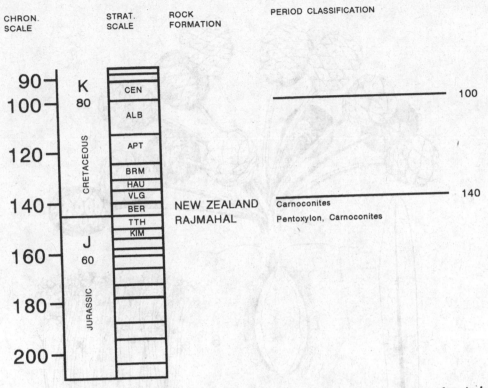

Fig. 8.35. The probable stratigraphic age of the two major southern hemisphere occurrences of the *Pentoxylon* group in Period Classification 140.

number were borne close together on a short branching axis. The assembly was made in one of the last papers by Birbal Sahni (1948). The pollen organ *Sahnia* was described by Vishnu-Mittre (1953) as a mass of oval 'sporangia' borne on a strangely forked axis (Fig. 8.36); the pollen was psilate monosulcate (Osborn *et al.* 1991), and not saccate. The Jurassic stratigraphy of the Rajmahal Hills has always been difficult to establish, but the descriptions by Harris (1962) of *Carnoconites* from the latest Jurassic–early Cretaceous rocks of New Zealand and by Drinnan and Chambers (1986) from Aptian rocks in Australia (Fig. 8.37) suggest that the date of the whole Indian discovery may prove to be of at least very late Jurassic age. Meyen (1987) repeated the suggestion that more than one plant was represented in the Indian material, but even he was attempting to establish evolution from the Carboniferous *Medullosa* group on the basis of morphology rather than of stratigraphy. Bose *et al.* (1985) in declining to categorise the *Pentoxylon* plant supported the view that this is one more of many Mesozoic plants that display an unexpected variety of form and demand a new and more open approach to classification.

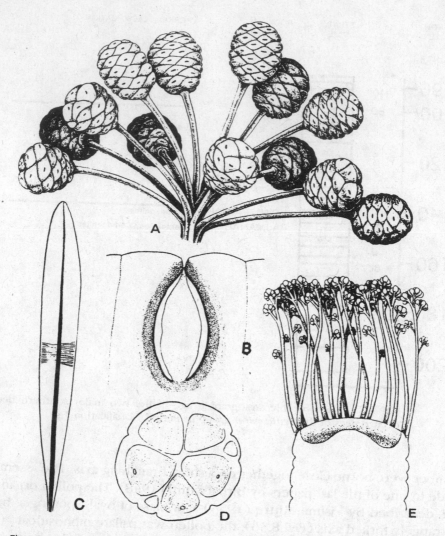

Fig. 8.36. The *Pentoxylon* group of plants from the southern hemisphere Late Jurassic: (A) *Carnoconites* ovulate heads, ×2; (B) section of ovule, ×7.5; (C) *Nipaniophyllum raoi* leaf, ×0.8; (D), *Pentoxylon* wood, transverse section, ×6; (E) *Sahnia* pollen-bearing flower, ×2 (from Crane 1985).

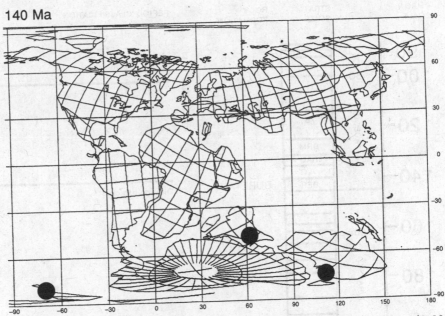

140 Ma

Fig. 8.37. Paleocontinental map for earliest Cretaceous Berriasian time, 140 Ma; plotted are areas of occurrence of fossils of the *Pentoxylon* group in eastern India, New Zealand and Australia (said to be Late Jurassic and Early Cretaceous) (map courtesy of A. G. Smith, 1992).

8.25–8.30 Cretaceous Berriasian–Valanginian records

8.25 Dirhopalostachys

This new group of fossils was outlined by Krassilov (1975), using material collected from Cretaceous Berriasian–Valanginian sediments of the Bureya River, Amur Basin of Far-East Asia (Fig. 8.38). *Dirhopalostachys* consists of a 5 cm slender axis with up to eight spirally arranged paired semi-closed carpels developing into one-seeded beaked cupules. The author believed the newly described plant to be of pro-angiosperm status similar to that of *Caytonia* and *Leptostrobus*, with which it occurred (Fig. 8.39). He also compared it morphologically with *Beania*, an ovulate structure of the Jurassic Nilssoniales and widely regarded as ancestral to living cycads; this comparison depended on association in the same beds with abundant *Nilssonia schmidtii*, a leaf relatively similar to Jurassic *N. compta* but differing in some cuticle characters. *Nilssonia schmidtii* was also stated to have some cuticle characters in common with *Dirhopalostachys*. However, the association argument was supported by elimination in that the three leaves *Sagenopteris, Phoenicopsis* and *Nilssonia* occurred with *Caytonia, Leptostrobus* and *Dirhopalostachys* as the only common elements in the beds concerned; of these the first two of each list were paired, but this does

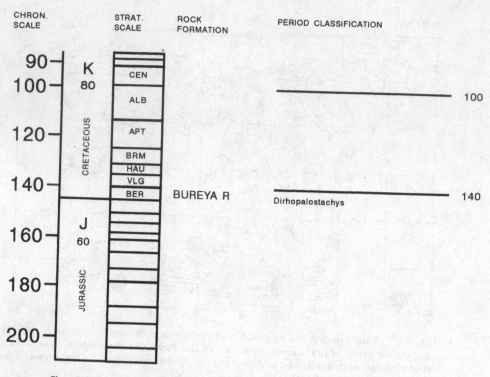

CHRON. SCALE | STRAT. SCALE | ROCK FORMATION | PERIOD CLASSIFICATION

Fig. 8.38. The stratigraphic age of the isolated Asian occurrences of *Dirhopalostachys* (Period Classification 140).

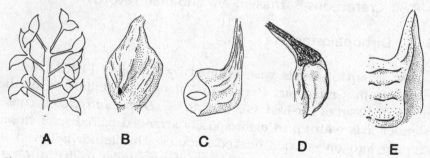

Fig. 8.39. Early Cretaceous *Dirhopalostachys*. (A) Reconstruction of gynoclad (ovuliferous fructification), ×0.75; (B) ventral aspect of single capsule, ×7.5; (C) transverse section; (D) ventral view dissected to show seed; (E) grooved beak of ripe capsule (from Krassilov 1975).

not conclusively prove the last of each list to be associated. It may therefore be preferable to regard *Dirhopalostachys* as a separate individual occurrence until confirmation of association is achieved in other strata; its own 'proangiosperm' status is clear.

8.26   Leptostrobus (Phoenicopsis) 3

See also Sections 8.10 and 8.23. Fossils of *Leptostrobus* and *Phoenicopsis*, but not *Czekanowskia*, are abundant in the Siberian Far-East early Cretaceous of the Lena and Amur Basins but all at more than 50° Cretaceous N (Fig. 8.40). In Cenomanian time and even through until the Early Senonian these plants were finally confined to the Okhotsk–Chukhotka region of the extreme Far-East at 60° Cretaceous N. Krassilov (1970) using well-preserved latest Jurassic material from the Far-East strengthened the links between *Phoenicopsis*, *Leptostrobus* and *Ixostrobus*; he also observed pollen grains among the papillate cells along

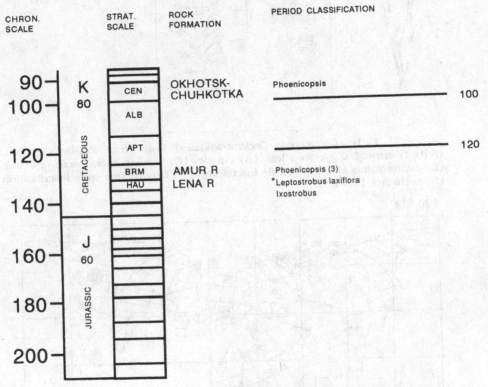

Fig. 8.40. The stratigraphic position of the Asian *Leptostrobus/Phoenicopsis* 3 occurrence on which pro-angiosperm features have been based (Period Classification 120).

the suture between the two valves of *Leptostrobus* capsules, suggesting that there was a possible form of stigmatic surface on the valve edge (Fig. 8.41). Harris (1976) pointed out that this did not apply to his English mid-Jurassic fossils; consequently there may even have been evolutionary development within the group in this direction during late Jurassic time (Fig. 8.42).

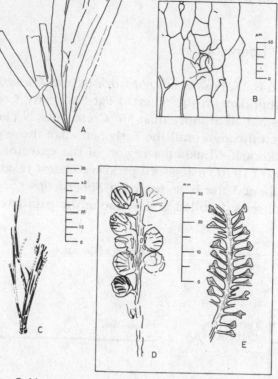

Fig. 8.41. Early Cretaceous Czekanowskiales from the Russian Far-East. (A,B) *Phoenicopsis speciosa* leaf; (B) cuticle. (C) *Czekanowskia aciculata* leaf; (D) *Leptostrobus laxiflora*, female fructification; (E) *Ixostrobus*, male fructification (from Hughes 1976).

120 Ma

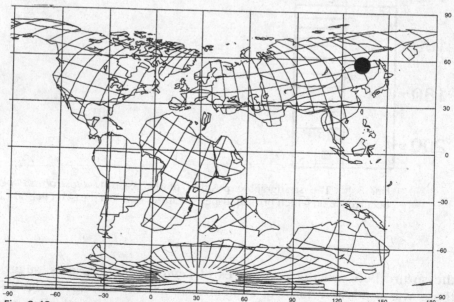

Fig. 8.42. Paleocontinental map for Cretaceous Early Aptian time, 120 Ma; plotted is the area of occurrence of *Leptostrobus* and *Phoenicopsis* in the Russian Far-East, paleolatitude 55° Cretaceous N; these fossils are no longer at this time known in any lower paleolatitude (map courtesy of A. G. Smith, 1992).

8.27   Sagenopteris *(Caytonia)* 3

See also Sections 8.13 and 8.19. Seward (1894–1895) described a number of leaflets from the Cretaceous Berriasian Ashdown Sands (Wealden) east of Hastings, Sussex, as *Sagenopteris mantelli* (Dunker); this species was originally erected for German Wealden material, among which it appears to have been more common (Fig. 8.43). No fructifications were known, but this discovery was thirty years before the first description of *Caytonia*. Similar records of *Sagenopteris* are known for Early Cretaceous rocks throughout Laurasia, but never on a very prominent scale. They also form part of the assemblage of relict Pollenifera groups including Czekanowskiales to be found only in the Okhotsk–Chukotka province of the Russian Far-East in Cenomanian times. These Cretaceous floras do not appear to provide any reproductive structures to match the *Sagenopteris* leaves, which means that it is not known whether a Cretaceous Valanginian cupule (when found) would display micropyles to its ovules or not. It should not be assumed therefore that Cretaceous fossils would precisely resemble Jurassic Bajocian *Caytonia* in this respect; in fact any assumptions of unchanged continuity, which are frequently made in the literature, are in practice denials of evolution. To find the ovular structure, which might well be quite different from *Caytonia*, to match these late *Sagenopteris* leaves (Fig. 8.44), is an important challenge.

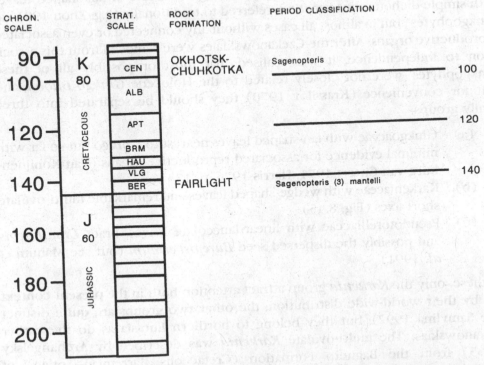

Fig. 8.43. The stratigraphic age of Cretaceous occurrences of *Sagenopteris* leaves in Europe and their continuation in Asia (Period Classification 140 and later).

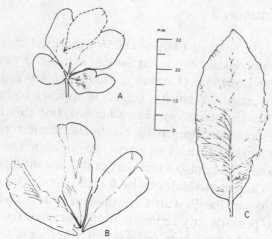

Fig. 8.44. Leaves of Early Cretaceous *Sagenopteris*. (A) *Sagenopteris mantelli*, Valanginian, Hastings, England; (B) *S. mantelli*, Aptian, Suifun Basin, Russian Far-East; (C) *Caytonia orientalis*, Aptian, Suifun Basin (from Hughes 1976).

8.28   The 'Karkenia' group

Throughout the Mesozoic era there were large numbers of fan-shaped leaves with simple dichotomous venation, referred to traditionally (e.g. Zhou 1991) as 'ginkgophytes', but in almost all cases without any connected or even associated reproductive organs. After the Czekanowskiales were removed from this association to independence, it was realised that most or possibly all of these 'ginkgophytes' were not closely related to the Holocene *Ginkgo biloba*, and that for convenience (Krassilov 1970) they should be separated into three family groups:

(a)   Ginkgoaceae with fan-shaped leaves nearest to *Ginkgo* but so far with minimal evidence for associated reproductive organs (Van Konijnenburg-van Cittert 1971, Harris 1976, p. 123).

(b)   Karkeniaceae with wedge-shaped leaves and remarkable multi-ovulate short axes (Fig. 8.45).

(c)   Pseudotorelliaceae with linear-lanceolate leaves, bract *Umaltolepis* and possibly the dispersed seed *Burejospermum* (but see Manum *et al.* 1991).

Of these, only the *Karkenia* group attract attention both in the present context and by their world-wide distribution; the other two groups are quite distinct (see Samylina 1992), but they belong to northern Eurasia as do the earlier czekanowskias. The multi-ovulate *Karkenia* was described by Archangelsky (1965) from the Baguero Formation (Cretaceous Barremian–Aptian) of Argentina together with a wedge-shaped leaf he named *Ginkgoites tigrensis* (Fig. 8.46); subsequently Krassilov (1970, 1972) found much more material

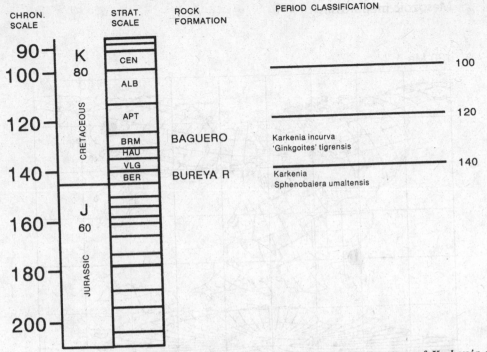

Fig. 8.45. The stratigraphic ages of two dispersed occurrences of *Karkenia* group fossils in Asia and in South America (Period Classifications 140 and 120, respectively).

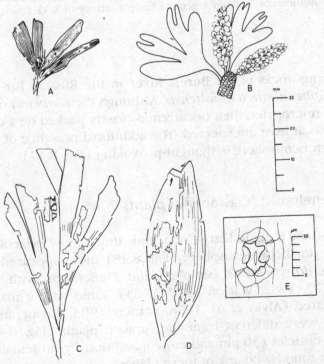

Fig. 8.46. Early Cretaceous fanleaves of the *Karkenia* group. (A,B) from Baguero Formation, Argentina; (A) *Ginkgoites tigrensis*; (B) *Karkenia incurva*, reconstruction; (C–E), from Russian Far-East; (C) *Sphenobaiera umaltensis* leaf; (D) *Eretmophyllum glandulosum*; (E) cuticle of *E. glandulosum* (from Hughes 1976).

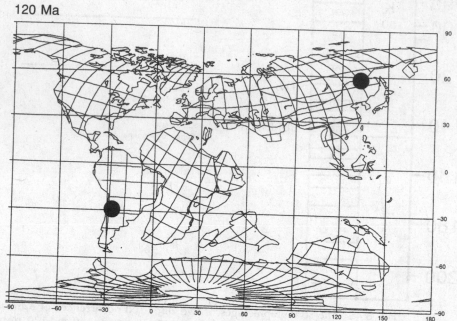

120 Ma

Fig. 8.47. Paleocontinental map for Cretaceous Early Aptian time, 120 Ma; plotted are occurrences of *Pseudotorellia* and *Karkenia* groups in Russian Far-East, and of *Karkenia* group unexpectedly in Argentina (map courtesy of A. G. Smith, 1992).

in the early Cretaceous rocks of the Bureja River in the Russian Far-East, together with the leaf *Sphenobaiera umaltensis*. Although the *Karkenia* ovules all appear to have short micropyles, their occurrence closely packed on a crowded unprotected spike was at least unexpected. The additional presence of this one group in the southern hemisphere is thought-provoking (Fig. 8.47).

8.29   Pseudofrenelopsis: 'Classopollis' plants 3

See also Sections 8.14 and 8.22. In Early Cretaceous time in low palaeolatitudes (Alvin 1982) in the northern hemisphere (Fig. 8.48) there appeared several species each of *Pseudofrenelopsis* (simpler) and *Frenelopsis* with unusual branching (Alvin 1977, 1982; Watson 1977, 1983), some with a male cone *Classostrobus* associated (Alvin *et al.* 1978, Francis 1983, Taylor and Alvin 1984). These plants were different from the Jurassic plants (Fig. 8.49) and possessed very thick cuticles (30 μm and more); additionally no female structures at all have been found with any of these plants.

Watson (1988), reviewing all the more significant and complete occurrences of 'Classopollis' plants from Late Triassic to Late Cretaceous time, referred them all to the single family Cheirolepidiaceae, which was then reasonable in view

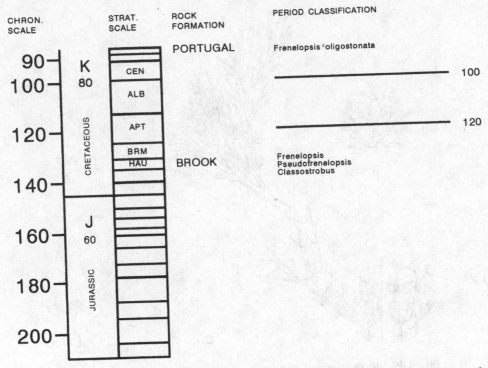

Fig. 8.48. The stratigraphic age of the probably separate *Classopollis* 3 plants *Pseudofrenelopsis* and *Frenelopsis* (entering in Period Classification 120).

of the tantalising lack of any knowledge on which a satisfactory division might have been based. The morphologic differences of the Cretaceous plants from the earlier ones may, however, be reflected in pollen distribution phenomena. Hughes (1973) suggested from (a) the dispersed pollen percentages (Early Cretaceous) rising steadily seawards down the river distributions, and (b) the progressively rising occurrence frequencies in lower paleolatitudes, that the plants occupied a subtropical sea-margin habitat later to be regarded as the mangrove zone. Originally this appeared to be in conflict with the views of Vakhrameev (1966, 1988), who was mainly concerned with Jurassic occurrences in Asia, but it now seems possible that different Cretaceous plants bearing different 'Classopollis' species occupied new niches and that both general views could be applied separately in different periods. The range of the dispersed pollen extends to Cretaceous Turonian time (Srivastava 1976); some later occurrences may have been reworked. A *Frenelopsis oligostomata* described from the Cretaceous Senonian of Portugal (Pons *et al.* 1980) may have existed slightly later. A southern hemisphere Early Cretaceous fossil *Tomaxiella biforme* from the Argentine Cretaceous Aptian Baguero Formation (Archangelsky 1968*b*), which has attached male and female organs and the thin cuticle of the northern hemisphere Jurassic fossils, was possibly a relict and isolated from the *Frenelopsis* occurrences in the Cretaceous N.

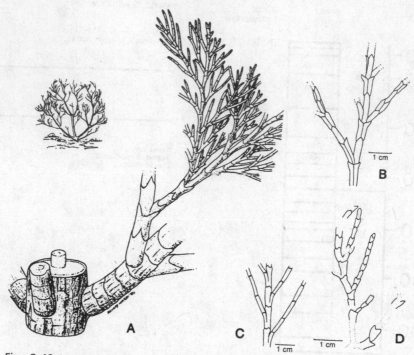

Fig. 8.49. Reconstruction and drawings of Early Cretaceous *'Classopollis'* plants. (A) *Frenelopsis ramosissima*, reconstruction of shrub, Aptian, USA, ×0.3; (B) *Cupressinocladus valdensis*, Valanginian, England, ×0.5; (C) *Frenelopsis hoheneggeri*, ×0.3 (D) *F. teineiriae*, ×0.5, both Hauterivian, western Europe (B–D, Alvin 1982; A, Watson 1988).

'Classopollis' 3 plants eventually succumbed to extinction in Late Cretaceous time. The argument that the possibly obligate tetrad at natural dispersal amounted to an angiospermid feature (Pocock *et al*. 1990), advantageous in reception on a stigmatic surface, is difficult to sustain unless it was unsuccessful or unless there was a Late Cretaceous actual angiosperm successor that has not yet been suggested. However, as pointed out by Harris (1979) and Alvin (1982), germination of pollen apparently took place outside the ovules and not necessarily through the micropylar beak, representing some kind of advance from the micropyle-drop mechanism.

8.30   Cycadeoidea 1

These fossils, normally permineralised in silica, consist of short, broad trunks covered in leaf bases and in some cases with flower buds. They were first described (Buckland 1828) from the Purbeck Beds (Jurassic Tithonian) of southern England, dated now as Cretaceous Berriasian (Allen and Wimbledon

1991); *C. gigantea* from these beds is 1.18 m high, but in this case without flowers. Although there are a small number of scattered records from early Jurassic rocks, the specimens are small and it is only in the earliest Cretaceous Purbeck Beds that records of large specimens begin. Some specimens correspondingly have also been recorded from the Morrison Formation in the United States. An important point concerning these fossils is that they are virtually indestructible, with the result that many of the best specimens have long been used by humans as garden or field furniture and their precise stratigraphic provenance has as a result been forgotten or never recorded. This applies among others to two of the best specimens in European museums: *Cycadeoidea reichenbachiana* in the Zwinger Museum, Dresden, was collected in 1753 from an unrecorded locality and described only in 1844 and 1892; *Cycadeoidea etrusca* Capellini et Solms 1892, now in Bologna, was taken from a 4000 year old Etruscan tomb. The original, quite rich, occurrence approximately at the Jurassic/Cretaceous boundary is separated in time by about 20 million years from the principal rich American locality and from some further European occurrences; see Section 8.35.

8.31–8.36 Cretaceous Hauterivian–Albian records

8.31 'Gnetophytes'–*Ephedripites*

Searching for ancestors of living *Ephedra, Welwitschia* and *Gnetum* in Mesozoic rocks is an honourable and necessary occupation (Krassilov and Bugdaeva 1988*b*), complementary to searching for ancestors of angiosperms and perhaps interwoven with it. It is not helped, however, by the weakly supported use in the Mesozoic of the terms 'gnetophyte' or 'chlamydosperm', which imply some single origin for all three plants. These living plants should be considered only as the end-product of diversification of some Mesozoic group or groups that have not yet been erected or named because the claims of finds are thus far so insubstantial. Van Konijnenburg-van Cittert (1992) bases such a hint on some Early Jurassic synangia with large polyplicate pollen, but other characters of her plant are quite diverse and give no support to this suggestion.

Leaving aside for the moment the isolated Triassic palynologic evidence (Pocock and Vasanthy 1988, Hughes 1992), *Drewria potomacensis* has recently been described (Crane and Upchurch 1987) from Potomac Zone I strata (probably Cretaceous Aptian) of Virginia as a small plant (6–7 cm high) with a sheathing leaf-base and leaves with reticulate venation resembling that of a *Welwitschia* cotyledon, plus dispersed pollen sacs with *Welwitschia*-type pollen in the same bed. This appears to be a most stretched (optimistic) comparison with a bizarre living plant of extremely isolated geographic and ecologic distribution, whose immediate late Cenozoic ancestors seem most unlikely to

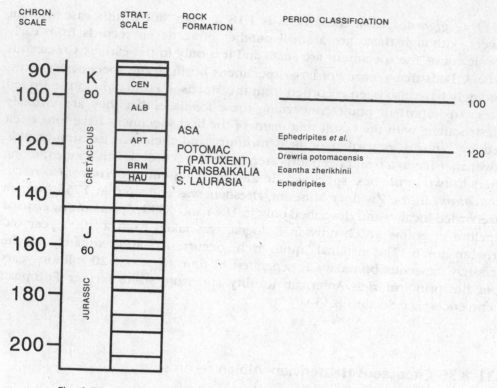

Fig. 8.50. The stratigraphic age of the short-lived Cretaceous influx of small species of *Ephedripites* in South Laurasia and Africa/South America (ASA), and of some associated megafossils (Period Classifications 120 and 100).

become paleontologically available; 100 million years is a long time to hold and to hide such diversity. Krassilov (1987) made a better case for *Eoantha zherikhinii* from the Early Cretaceous of Transbaikalia; he (Krassilov 1986) described as an ovulate organ with four cupules each containing an orthotropous ovule with *Ephedripites*-type pollen grains in the pollen chamber. Even this case, for its approach to 'gnetophytes' depends solely on the pollen *Ephedripites*, which appears suddenly in quantity as a rather small dispersed grain (<30 μm) in latest Hauterivian time in South Laurasia (Fig. 8.50); this appearance closely parallels the entry of the monosulcate columellate–tectates (MCT group) in the same region (see both Section 8.33 and Chapter 9). The polyplicate pollen *Ephedripites* has of course only one analogue in living floras, hence the derivation of its name, which is also apparently nomenclaturally insecure; the diversity already recorded in the Cretaceous dispersed pollen (Hughes 1992), considerably exceeds in morphology and size range that in the species of living *Ephedra* (see Figs. 9.29 and 9.30 below). In the Africa–South America (ASA) province from Aptian to Early Cenomanian time much greater diversity of such pollen has been reported (see Herngreen and Duenas Jiminez 1990, Osborn *et al.* 1993), together with unique presence of the elaterate group. The dispersed

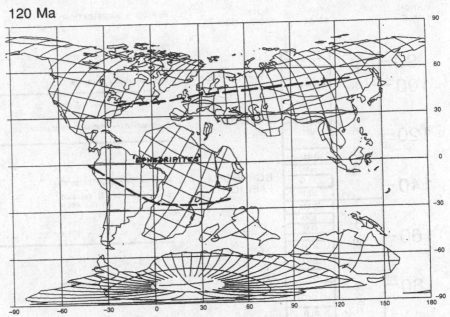

Fig. 8.51. Paleocontinental map for Cretaceous Early Aptian time, 120 Ma; distribution of occurrences of early Cretaceous *Ephedripites* pollen of unknown affinity (map courtesy of A. G. Smith, 1992).

polyplicate pollen (Fig. 8.51) remains common in both regions mentioned above only until Cretaceous Cenomanian time, and need not have any direct connection with the widespread warm-temperate arid zone distribution of living *Ephedra*, which has only a few late Cenozoic megafossil records and does not lend itself either in morphology or in distribution to distinctive unequivocal fossilisation. Finally the entirely different living *Gnetum*, with numerous species of rainforest tree or liana, has a very simple, small pollen grain with a distal leptoma; Krassilov (1987) in passing compared the detail of the distal pore of *Classopollis*, which comparison keeps extensive possibilities open but demands no immediate action. Some good early or mid-Tertiary fossils would stimulate this discussion.

8.32   Eucommiidites *plant 2*

The dispersed pollen *E. troedssonii* (diameter 30–40 μm) continues into Cretaceous time, although with declining frequency (Fig. 8.52); see also Section 8.17. A smaller species *E. delcourtii* (diameter about 25 μm, with relatively thicker exine) was found (Hughes 1961) in Cretaceous Valanginian ovules with long micropyles from Sussex, England (Fig. 8.53), and also in similar ovules (Brenner 1967) probably of Cretaceous Aptian age from the Potomac Group, eastern

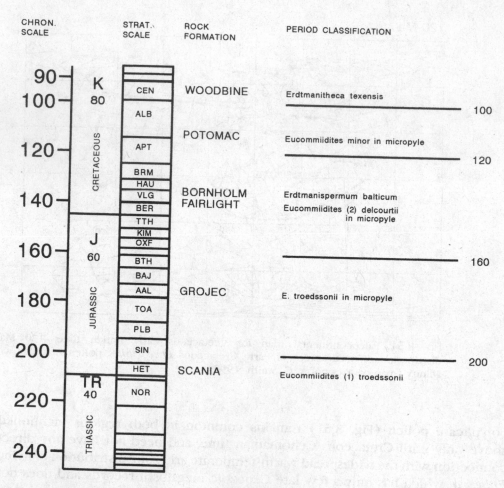

CHRON. SCALE | STRAT. SCALE | ROCK FORMATION | PERIOD CLASSIFICATION

Fig. 8.52. The stratigraphic distribution of occurrences of what may be called the *Eucommiidites* plants of Jurassic and Cretaceous rocks of South Laurasia (Period Classifications 200 to 90 inclusive).

USA. This limited mesofossil information has recently (Pedersen *et al.* 1989) been supplemented from the Cretaceous Cenomanian rocks of the Woodbine Formation of Texas; *Erdtmanitheca texensis* consists of spherical heads of about 1 cm diameter with radiating scales and pollen sacs containing smooth small *Eucommiidites* pollen of 24 μm diameter. Pedersen *et al.* (1989) also record *Erdtmanispermum balticum* from Cretaceous Valanginian rocks of Bornholm, Denmark; the seed (just over 1 mm long) with an elongated micropylar tube containing small *Eucommiidites* pollen (22 μm in diameter) said to resemble those from England (Hughes 1961). As is evident from the above discussion, the vegetative and other main features of the plant concerned are not yet known; the long micropyles have been suggested to resemble those of living *Ephedra* species but there appears to be no other reason to support such affinity.

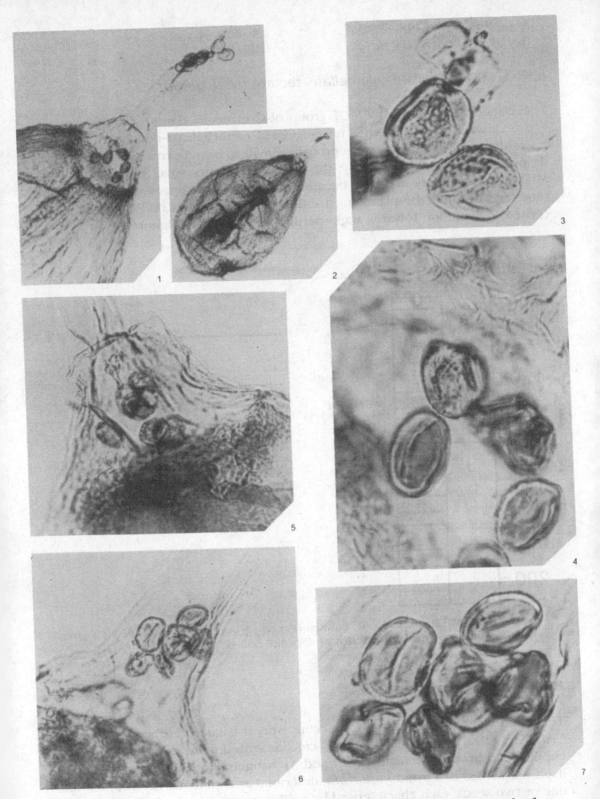

Fig. 8.53. Photomicrographs of Early Cretaceous *Eucommiidites* in the micropyle of an ovule. (1,2,5,6) *Spermatites pettensis*, Valanginian, southern England, ×45 (Fig. 2); (3,4,7) *Eucommiidites delcourtii*, within the micropyle, ×1200 (from Hughes 1961).

8.33   *Monosulcate columellate–tectate (MCT group)*

The dispersed pollen of the MCT group of Cretaceous Hauterivian–Barremian age is described in Chapter 9; no megafossil of the plants concerned has yet been identified. The plants were probably small to have been so far overlooked despite their probable diversity; the pollen is all very small (Fig. 8.54). The MCT group is mentioned here because of its undoubted presence at a critical time, and because the plants must either be angiosperms or Pollenifera concerned in evolution towards angiospermy; they cannot be ignored.

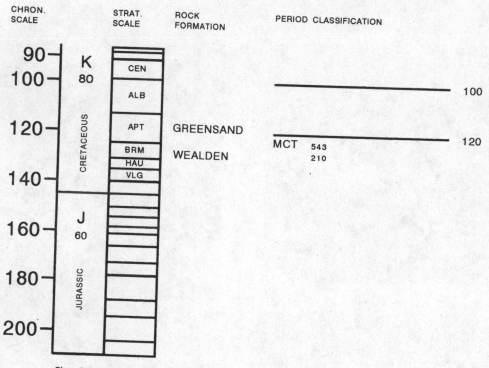

Fig. 8.54. The stratigraphic position (Period Classification 120) of the MCT group of unknown plants as indicated by dispersed pollen.

8.34   Ktalenia

This fossil (Fig. 8.55) from the Cretaceous Aptian Baguero Formation of Santa Cruz, Argentina, consists of a slender 2 cm axis with downward facing spherical cupules and bracts attached (Taylor and Archangelsky 1985). The fertile axis itself is attached to a frond *Ruflorinia*, described as tripinnate. Each cupule has one or two seeds, each characterised by a short nucellar beak; the fruit appears to be indehiscent and relatively similar to that of *Caytonia*, although the leaf is different (Fig. 8.56). The preservation is favourable enough to provide good

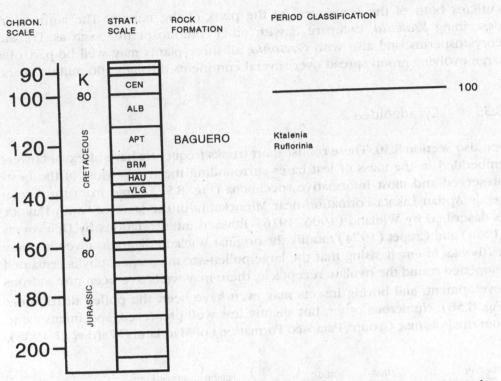

STRAT.
SCALE

ROCK
FORMATION

PERIOD CLASSIFICATION

Fig. 8.55. The stratigraphic position of the isolated Cretaceous southern hemisphere plant *Ktalenia* with some resemblance to *Caytonia* (Period Classification 100).

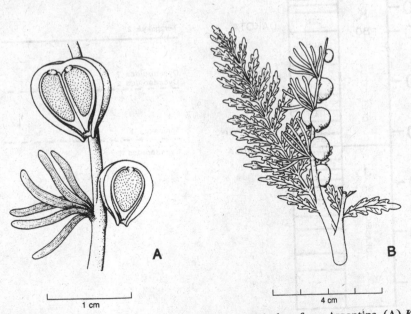

Fig. 8.56. Early Cretaceous cupule-bearing seed plant from Argentina. (A) *Ktalenia*, portion of fertile frond with sectioned cupules; (B) reconstruction of *Ruflorinia* frond and *Ktalenia* axis (from Taylor and Archangelsky 1985).

cuticles both of the leaves and of the parts of the cupule. The authors, in describing *Ktalenia*, compare it with other 'pteridosperms' such as Triassic corystosperms, and also with *Caytonia*; all three plants may well be part of a large evolving group spread over several continents through 100 million years.

8.35   Cycadeoidea 2

See also Section 8.30. These robust short trunks frequently have bisexual flowers embedded in the mass of leaf bases surrounding the trunk. Most of the best-preserved and most informative specimens (Fig. 8.57) came from the Cretaceous Aptian Lakota Formation near Minnekahta in the state of South Dakota, as described by Wieland (1906, 1916). Revised interpretations by Delevoryas (1968) and Crepet (1974) modify the original Wieland diagrams given in many textbooks in emphasising that the large pollen-sac microsporophylls remained unopened round the ovulate receptacle; there may well have been protandrous development, and boring insects may even have been the pollen distributors (Fig. 8.58). Numerous other, but slightly less well preserved, specimens came from the Potomac Group (Patapsco Formation) of Maryland (Ward *et al.* 1905),

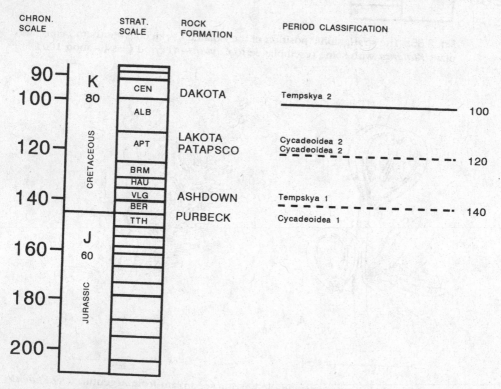

Fig. 8.57. The stratigraphic ages of the two Cretaceous radiations of *Cycadeoidea* and of *Tempskya* (Period Classifications 140 to 90, inclusive).

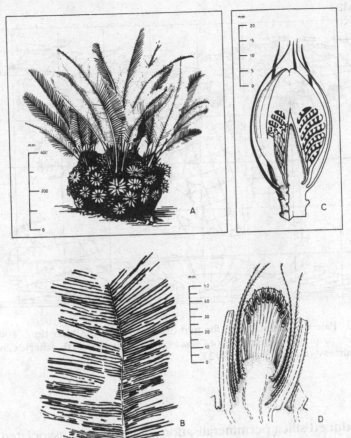

Fig. 8.58. Early Cretaceous *Cycadeoidea* plants: (A) reconstruction of *Cycadeoidea marshiana*, × 1/20; (B) *Pseudocycas saportae*, upper third of frond, ×0.4; (C) reconstruction of single protandrous bisexual flower, ×0.75; (D) *C. wielandii*, flower with mature ovules, ×0.4 (from Hughes 1976).

where because of the unconsolidated strata and low dip the uncertainty of stratigraphic provenance is even more acute (Fig. 8.59). A number of specimens were also collected in the last century (but seldom since) from Cretaceous Aptian rocks of western Europe, including England. One of several of these from the Isle of Wight provided Wieland (1934) with sections of a unique petrified crown of buds that implied that the leaves were stiff and pinnate, perhaps of the *Pseudocycas* type, which is a common Early Cretaceous compression fossil. This and the other excellent reconstructions, mostly based on South Dakota material, conceal the perennial difficulty that permineralisation and compression fossils can never occur together because of the necessarily different sedimentation conditions.

Despite the durability of *Cycadeoidea* fossils, it is uncertain whether their range of occurrence in time and/or space is merely a product of the environment

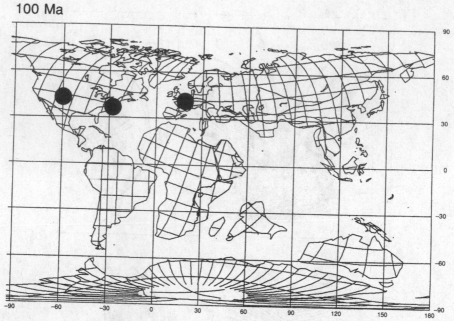

100 Ma

Fig. 8.59. Paleocontinental map for mid-Cretaceous Late Albian time, 100 Ma ago; plotted are principal occurrences of *Cycadeoidea* group in North America and Europe (map courtesy of A. G. Smith, 1992).

potential that produced silica permineralisation and is usually associated geologically with volcaniclastic facies. Further widely scattered records have been reported from western Poland, European Russia, Mongolia and Japan.

Tempskya. In the context of mineralisation preservation, it is strange that a comparably preserved stem *Tempskya* (described as a tree-fern) should be found very widely distributed (Ash and Read 1976) from Montana and Idaho through to New Mexico in the mid-western USA in rocks of the Dakota Sandstone of Cretaceous Albian age (Fig. 8.60), i.e. a little later than the main *Cycadeoidea* occurrence. Further, the original record of *Tempskya* was made under the name *Endogenites* by Fitton (1836) from Cretaceous Berriasian strata near Hastings, England, a similarly short time later than the Purbeck Cycadeoidea. Thus, as represented by two quite separate fossils, the special preservation conditions existed in the western half of Southern Laurasia for most of Early Cretaceous time, and is indeed supported by occasional other records in this region. This curious dual distribution in which the plants do not occur together may imply that their apparent ranges of occurrence are genuine, and not purely a taphonomic accident.

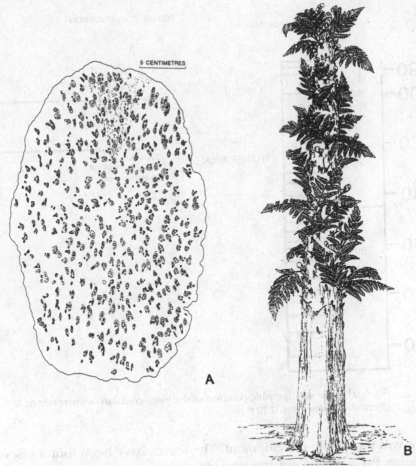

Fig. 8.60. The Early Cretaceous plant *Tempskya*: (A) *Tempskya reesidei*, sketch of cross-section of large permineralised false-trunk, stems circular but elongated near margin, roots drawn only in small sector (at the top of the figure); (B) reconstruction (2 m high) based solely on cross-section of *T. wesselii* by Andrews and Kern (1947) (from Ash and Read 1976).

8.36   Semionogyna

This was described (Krassilov and Bugdaeva 1988) as a protocycadopsid pteri-dosperm, which perhaps suggests a healthy degree of uncertainty. The fossils came from 'Neocomian–Aptian' rocks of the Semion Valley, 50 km south of Chita in Transbaikalia (Fig. 8.61), and consisted of leaves of *Cladophlebidium* (a bipinnate frond), *Semionogyna bracteata* (an 8 cm ovulate spike) and *Semionandra laxa* (a loose cone with persistent bracts and pendent sporangia). The three organs were associated and had adequate features in common but were not proved to belong to one plant. The pollen was described as anasulcate, and of 30 µm length. The ovules were orthotropous and the stomata anomo-cytic, and the authors favoured a cycad connection rather than pteridosperm

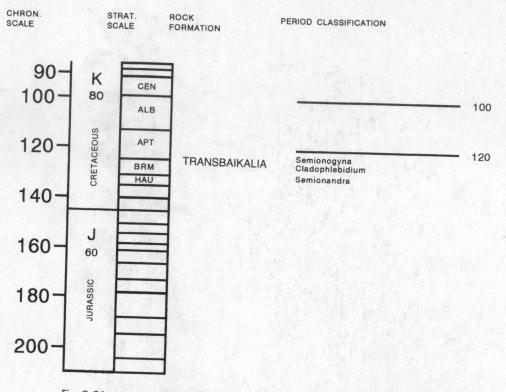

Fig. 8.61. The stratigraphic position of the isolated Asian occurrence of *Semionogyna* (Period Classification 120).

as there was no cupule development. The leaves have been found elsewhere in the Russian Far-East but not the reproductive organs. Such a group of fossils is not classifiable and illustrates the need to keep all options open if interpretations friendly to the ultimate understanding of evolution are desired.

8.37   Ancestors for living gymnosperm groups

The total number of living gymnosperm species (750) is very small indeed when compared with that of living angiosperms, and small even when compared with the Mesozoic seed-plants (Pollenifera). It is unnecessary therefore to set aside many Mesozoic plants as probable ancestors for this small number of extant gymnosperms and I suggest that only the following be so considered until other lineages are positively demonstrated:

(a)   A small group around *Nilssonia/Beania/Androstrobus* to lead to the living *Cycas, Zamia* and eight other genera (but see Krassilov and Bugdaeva 1988*a*).

(b)   A few *Ginkgoites* leaves that have minimal supporting reproductive organs to lead to the single living species *Ginkgo biloba*, but there is no adequate excuse for using the actual genus *Ginkgo* for any

Jurassic fossil (see Tralau 1967) or for any Cretaceous fossil (see Drinnan and Chambers 1986).

(c) Nothing as yet for any of *Gnetum, Ephedra* or *Welwitschia*, despite the efforts to this end of Crane (1988) and Pocock and Vasanthy (1988).

(d) The problem with the main group of 700 conifers is different because the majority of all Mesozoic seed-plants have simple small or short leaves with thick cuticles and some kind of cone reproductive structure. It has been only too easy to refer most of these optimistically to one or other living conifer family. However, most serious authors admit that the characters involved in Mesozoic examples never quite fit the living families even with the few organs usually available (Harris 1979, Fisher and Watson 1983, Krassilov 1987, Bose and Manum 1990, Tidwell and Medlyn 1992), and consequently it is doubtful whether any such attribution of Jurassic fossils advances knowledge at all. The occurrence of *Classopollis*-like pollen in a succession of such plants differing through Jurassic and Cretaceous time (Watson 1988) and many of them outwardly similar to species bearing more usual types of pollen, emphasises the shortage of knowledge so far and the futility of premature classifications which have been wrongly allowed to influence datasets as well as nomenclature. The best-documented cases for attribution include Early Cretaceous fossils for the Pinaceae on the basis of details of cones (Alvin 1957), but even this evidence could perhaps be best included in a separate Cretaceous group on an 'evolutionary pathway' rather than be squeezed into an extant family; the latter process ends all thought about what might have happened in the last 150 million years.

8.38   Status of Mesozoic 'conifers'

As mentioned above the possession in the Mesozoic of either brachyphyll or linearphyll leaves with a thick cuticle provides no certainty that the fossils were direct ancestors to any living conifer family (e.g. Watson *et al.* 1988). Some groups of fossils have already been shown to have other characters inappropriate for conifers and have been removed; further cases may well follow. The exacting criteria for evidence of evolution towards angiospermy should be parallelled by similarly stringent criteria for the origin of Pinaceae, Taxodiaceae or Araucariaceae. Taxa not shown to be involved directly in ancestry of living conifers should remain completely unaffiliated, which is one purpose of the Period Classification scheme.

8.39   Megafossil conclusion

It is clear that megafossils occur so widely separated both in time and space that single occurrences can be of only very restricted significance whatever

SOME ANGIOSPERMOUS FOSSILS

ANGIOSPERMS · DICOT · GYMN 2 · GYMN 1

Time scale (Ma): 0 · 20 · 40 · 60 · 80 · 100 · 120 · 140 · 160 · 180 · 200 · 220 · 240

Cyc. Zam. · Gnet. · Eph. Gink. Cupr. Pinac.

Labels:
Sag. · Pseudot. · Class. 3
Cycoid. 2 · MCT · MCT · Semio. · Ktal. · MCT · Kark. · 'Eph' · Eucom. 2 · Class. 3
Cycoid. 1 · Will. 3 · Lepto. 3 · Dirhp. / Sag. 3 · Pentox. · Kark. · Pseudot. · Class. 3 · Fan. · Brach. · Lin.
Nilss. · Wiella. · Will. 2 · Pachy. · Lepto. 2 · Cayt. 2 · Class. 2 · Fan. · Brach. · Lin.
Iran · Eucom 1
Will. 1 · Lepto. 1 · Pelt. 2 · Cayt. 1 · Sanmg. · Dinop. · Class. 1
Coryst. · Pelt. 1 · Gloss. 2

morphologic surprises they may provide. For example the well-preserved mid-Jurassic material of *Williamsoniella* falls into the Period Classification 160 and is as far in time from a Cretaceous Santonian fossil of still uncertain affinity as the latter is from the Holocene classification. Until substantial successor material for *Williamsoniella* becomes available in the Period Classifications 140 and/or 120, further speculation appears to be unprofitable.

World-wide search is specially needed for new megafossil floras, or failing that for mesofossils, in Cretaceous Berriasian–Barremian rocks. They are urgently needed both to provide linking successors to well-known Jurassic and Triassic fossils, and particularly to elucidate the provenance of the Hauterivian–Barremian tectate monosulcates; the latter are perhaps best sought narrowly in late Barremian time (MCT Phase 4; see Chapter 9) when the dispersed pollen becomes adequately represented. As already successfully demonstrated (see e.g. Dilcher and Crane 1984, Crane and Dilcher 1984) for the Albian tricolpate dispersed pollen, flowers can be discovered even if they are very small and devoid of distinctive corolla. A diagram has been constructed (Fig. 8.62) to express a general view that the majority of the taxa of the Pollenifera of Early Cretaceous times were candidates for further evolutionary development. Most of them had advanced characters of some kind, and an open polyphyletic approach to the problem is desirable to stimulate search for phenomena that might otherwise be relegated as botanically unlikely. Such is the diversity of Holocene seed-plants that it is unhelpful to constrain thought by prejudices about systematics when only new fossil discoveries will ever move the problem towards solution.

Fig. 8.62. General diagram to illustrate the distribution in time of the principal groups of Mesozoic Pollenifera, and of their representation in the successive Period Classifications from 200 to 120 Ma (geologic time-scale in millions of years ago). The shrinkage to two dimensions on paper, with one necessarily devoted to geologic time, precludes expression of some supposed proximities of relationship between fossil groups and with the much later extant Holocene phenomena. An underline of a name on the diagram itself indicates the main and sometimes only significant source of information on the group concerned. Abbreviations employed for fossils: Brach., Brachyphyll group (Hughes 1976); Cayt., *Caytonia* 1 and 2; Class., *Classopollis* plants (1 and 2 *Hirmeriella*; 3 *Frenelopsis*); Coryst., Corystosperm group; Cycoid., *Cycadeoidea* 1 and 2; Dinop., *Dinophyton*; Dirhp., *Dirhopalostachys*; Eph., *Ephedripites* group (without known affinity); Eucom., *Eucommiidites* 1 and 2; Fan., Fanleaf group; Gloss. 2, Late Permian *Glossopteris*; Iran., *Irania hermaphroditica*; Kark., *Karkenia* group; Ktal., *Ktalenia*; Lepto., *Leptostrobus* group (1 and 2 Czekanowskiales, 3 *Phoenicopsis*); Lin., Linearphyll group (Hughes 1976); MCT., monosulcate columellate–tectate pollen plants; Nilss., Nilssoniales; *Pachy, Pachypteris*; Pelt., Peltasperms 1 and 2; *Pentox., Pentoxylon* group; Pseudot., *Pseudotorellia* group; Sag., *Sagenopteris* 3 (leaf of *Caytonia*); Sanmg., *Sanmiguelia lewisii*; Semio., *Semionogyna* plant; Wiella., *Williamsoniella*; Will., *Williamsonia* 1,2 and 3. Abbreviations employed for extant plants: Cupr., Cupressaceae; Cyc., Cycas group; 'Eph'., *Ephedra* and *Welwitschia*; Gink., *Ginkgo biloba*; Gnet., Gnetum only; Pinac., all conifers other than Cupressaceae; Zam., Zamia group.

9 English Barremian monosulcate pollen

9.1 Research project

Over the last fifteen years, extensive records have been made of apparently angiospermid pollen from Cretaceous Hauterivian, Barremian and Aptian rocks of southern and eastern England (Fig. 9.1). It is believed that the total available English rock succession containing such pollen is more complete than that of any other area so far worked. The largely non-marine strata represent relatively continuous deposition through from Late Jurassic to mid-Cretaceous time in southern England; the semi-marine and marine successions in the north of the region help to guide stratigraphic correlation and provide a link with the nearby Global Stratigraphic Scale points in western Europe. The outcrops fall now in a cool temperate area with minimal destructive weathering problems; the cored borehole back-up presents some but not too many problems of interpretation. There are very few tectonic or structural complications in the main successions of the strata and, although the outcrops are now disjunct, correlations already made are acceptable, as also are correlations with further suitable rocks in northern France, northern Germany and elsewhere. Because the pollen grains are all very small, they have been strew-searched on gridded stubs by SEM; because in the lower parts of the successions it was necessary to search downwards to 'failure of occurrence', with the frequencies very low (and the search time per grain correspondingly long), observation for record has so far been restricted to SEM, leaving most of the appropriate TEM studies for the future.

Interest in these English occurrences began with early light microscope work (Couper 1958, Hughes 1961, Kemp 1968), but intensive study and much better understanding had to await the scanning electron microscope and successful handling of very small single dispersed fossil grains. Parallel work in the eastern United States (Brenner 1963, Doyle 1973, 1983, Upchurch 1984, Walker 1974, Walker and Walker 1984) on the Potomac Formation has also yielded valuable results, but in the earlier parts of that succession with more difficulty from less favourably stratified successions.

The fossils here recorded are predominantly the monosulcate columellate–tectate or semi-tectate pollen of this age, together with numerous other types of associated and possibly relevant grains. All of these pre-date the more obvious general mid-Cretaceous (Albian onwards) arrival of tricolpate and other multi-

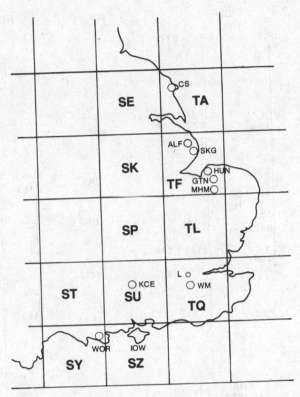

Fig. 9.1. Map of locations of Hauterivian–Aptian successions studied in southern and eastern England, marked with National Grid Reference squares. WM, Warlingham Borehole in Surrey; KCE, Kingsclere Borehole in Berkshire; WOR, Worbarrow Bay in Dorset; IOW, Isle of Wight; HUN, GTN, MHM, Hunstanton, Gayton, Marham Boreholes in Norfolk; ALF, SKG, Alford, Skegness Boreholes in Lincolnshire; CS, Speeton in East Yorkshire. L, London, for reference.

aperturate pollen. The various kinds of monosulcate pollen are not as yet associated with any convincing megafossil evidence of plants from which the pollen might have been derived. Comparison of this pollen with that of certain selected extant plants may stimulate the search for descriptive characters but is not here considered in any respect as evidence of affinity. It is not yet clear whether the various types of pollen and the providing plants, when ultimately revealed, may be considered as angiosperms or not, but they clearly form an important part of the story of evolution of the Pollenifera (seed-plants of the Mesozoic).

9.2   The geologic circumstances

From having been a general area of marine deposition in Jurassic up to Kimmeridgian time (see Harland *et al.* 1990 for time-scale), southern England received restricted marine (Portland Beds), then some evaporites (Purbeck Beds),

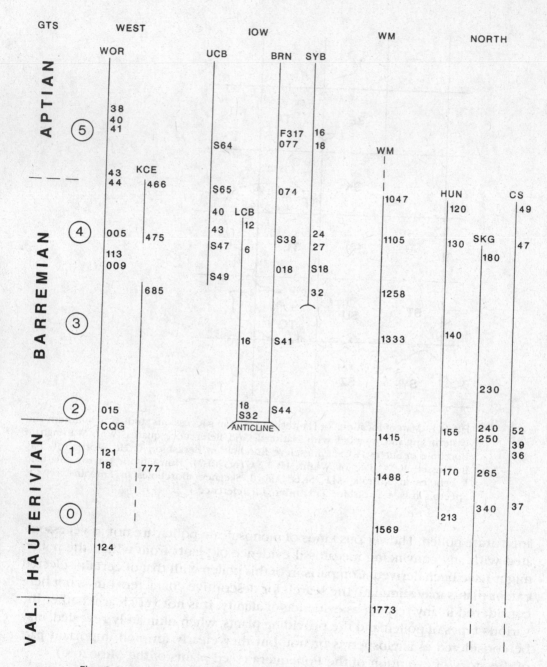

Fig. 9.2. Comparative table of the ten main successions employed. Some sample numbers are given with each column but see Figs. 9.3–9.7 for further details. Full line, succession available and adequately sampled; dashed line, succession available. WOR, Worbarrow; KCE, Kingsclere Borehole; IOW, Isle of Wight; WM, Warlingham Borehole; HUN, Hunstanton Borehole; SKG, Skegness Borehole; CS, Speeton. The scale is approximate geologic time and not rock thickness.

through to essentially non-marine sediments (Wealden Beds up to 600 m thick) in Jurassic Tithonian to Cretaceous Barremian time. Relative uplift of the London area and northward led to non-deposition there, erosion and contribution of sediment to the southern area. Any connections intermittently of the Wealden Basin to the sea were principally to the south across France to Tethys, and possibly westward. From Cretaceous Berriasian time onwards, a Boreal Sea (marine deposition), shared with northern Germany, the North Sea (to be) area and beyond, left deposits in East Yorkshire, Lincolnshire and Norfolk. In Cretaceous Aptian time a marine connection was formed from Tethys through France and southern England to the Boreal Sea. In Albian times this connection increased and became a widespread north-west Europe sea, maintained and enlarged as the Chalk Sea for the remainder of the Cretaceous period. The local proximity, or otherwise, of the individual succession areas to contemporaneous land affected the palynomorph assemblages separately in each case. Single geologic successions that illuminate fully all the fossils and paleoenvironments of a selected time-slice are very seldom available, and in detail Early Cretaceous southern England is no exception to this difficulty. All twelve of the successions mentioned below (Fig. 9.2) contribute something of unique significance that is elusive or unconfirmed in the others.

9.3   *The individual rock successions (south)*

In each succession the correlation with the Global Time-Scale (GTS) is shown with as much clarity as it is known. The boreholes were all logged in feet and inches, and the original sample numbers depend on this and have been retained; a depth scale in metres is also shown. All successful or significant samples are entered on the diagrams for reference. Outcrop sample numbers are field numbers, sometimes given on several separate visits and thus they display no logical arrangement.

Warlingham Borehole. The British Geological Survey made this valuable cored exploration hole (TQ 3476 5719) just south of London in 1954, to elucidate the presumed northward subsurface thinning against the 'London Ridge' of the thick Wealden deposits of Berriasian–Barremian age of the main basin of the Weald (Worssam and Ivimey-Cook 1971). Fortunately the thinning was much less than expected and the succession provides over 100 m (350 feet) of what may well represent the whole of the Barremian interval (Fig. 9.3), with good palynomorph retrieval. The Aptian strata above were less fully cored and more strongly marine in Lower Greensand glauconitic facies, and so provide for these purposes a less than satisfactory continuation. Below the Barremian, the first 30 m of Hauterivian strata are valuable, but downwards there is some confusion due to small-scale faulting, some repetition and probably omission, and some less suitable lithology; much lower, the Valanginian and Berriasian ages were

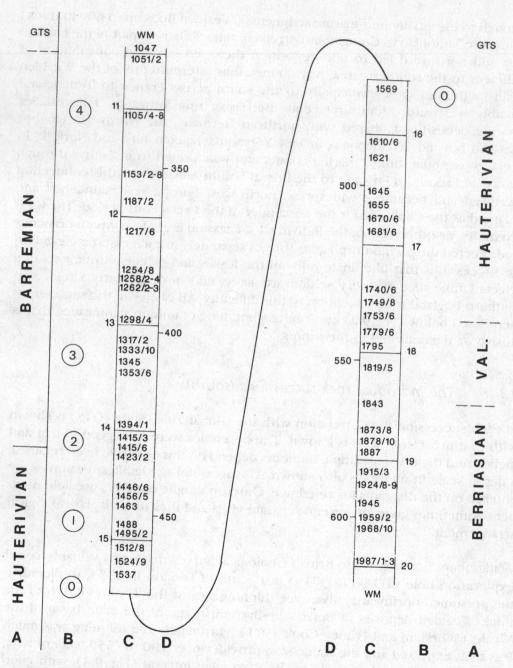

Fig. 9.3. Record of the most successful palynologic samples from the Wealden formation of the Warlingham Borehole (WM). A, ages of the geologic time scale (GTS); B, Phases 0–4 of the MCT assemblages; C, rock sample numbers that were originally borehole depths in feet; D, borehole depths in metres.

well represented and rich in palynomorphs (see Hughes and Moody-Stuart 1969). Other than palynomorphs including a few scattered dinoflagellates, the fossils available for correlation are principally ostracods, which are absent or unhelpful at many levels between the Berriasian beginning and the Aptian beginning taken at WM1050; the Hauterivian/Barremian boundary (Harding 1990) is believed to be accurately placed from dinocyst evidence but other boundaries are less certain.

Isle of Wight. This large island (20 miles × 10 miles (1 mile = 1.6 km)) provides four, excellent, well-exposed coastal successions of Barremian age, in three of which the upward continuation into Aptian strata is equally well exposed, of a less fully marine facies than Warlingham, and consequently richer in pollen. Unfortunately the earliest Barremian and all earlier strata are unexposed, being cut off from study at an anticlinal structure (Fig. 9.4A, B); there are no adjacent cored boreholes. The main 'Atherfield' Barremian–Aptian succession runs eastward, with a low dip from the anticline at Hanover Point (SZ380 850 on the southwest coast) for about seven miles to Atherfield Point, Blackgang Chine and Rocken End, where fully marine Late Aptian strata outcrop. Some parts of the 150 m thickness of Barremian strata are barren because they were oxidised at or soon after deposition, but there are many rewarding samples throughout. Treated here separately the 'Lower Compton Bay' succession (Fig. 9.4A), running northwestward from the Hanover Point anticline with a higher dip, presents most of the Barremian but not the Aptian rocks before terminating at an east–west fault. North of the fault, the 'Upper Compton Bay' succession presents rocks from the repeated mid-Barremian conformably through Aptian and Albian strata into Cenomanian and higher chalks near Freshwater; monosulcate pollen has been recorded from the Barremian and Aptian sections.

Finally and quite separately eight miles away Barremian–Aptian rocks (Fig. 9.4B) are well exposed on the east coast of the island north of Sandown in the 'Sandown–Yaverland' succession; once again everything before mid-Barremian is lost to sight (exposure) in the Sandown anticline. The coastal erosion of these fine sections is so rapid that published descriptions of the strata are mostly out of date in detail and good collecting localities will change as many of the favourable strata were formed in lenses. This is, of course, a desirable situation for new samples with unexplored potential.

Worbarrow Bay. Thirty miles west of the Isle of Wight, the western parts of the Wessex Basin in Dorset County present numerous coastal exposures from Swanage to west of Lulworth. Most of these show high dip and are incomplete from structural disturbance in too close proximity to the Purbeck anticline; only the section in Worbarrow Bay (SY 870796, approach from Tyneham), east of Lulworth and in a military reservation, is complete and tectonically straightforward to interpret.

The succession is continuous from late Jurassic Portland and Purbeck Beds,

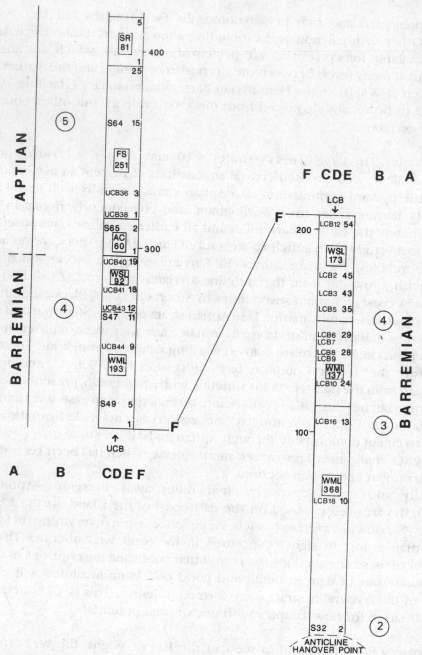

Fig. 9.4 (left). Record of the more successful palynologic samples from the Wealden Formation of the faulted succession in Compton Bay, Isle of Wight. A, Ages of the geologic time-scale (GTS); B, Phases 2–5 of the MCT assemblages; C, sample numbers; D, rock unit thicknesses in feet (White 1921); E, bed numbers (White 1921); F, approximate rock thickness in metres measured from base of succession at Hanover Point.

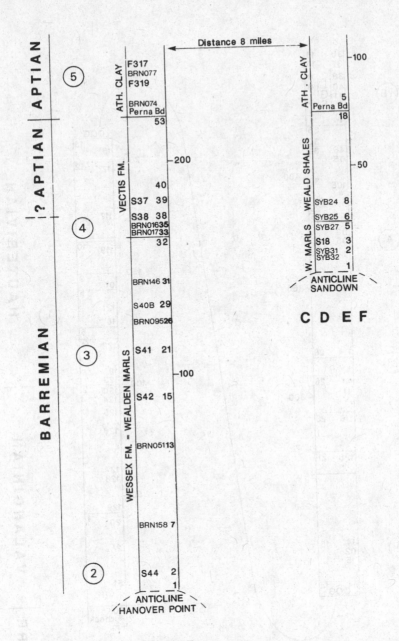

Fig. 9.4 (right). Record of the more successful samples from the Wealden Formation of the main Isle of Wight succession from Hanover Point (west) to Atherfield Point and Rocken End (east). A, ages of geologic time-scale (GTS); B, Phases 2–5 MCT assemblages; C, present names of formations; D, sample numbers; E, bed numbers (White 1921); F, approximate rock thickness in metres measured from base of succession. Right column, succession at Redcliff, north-east of Sandown; palynologic correlation remains unsatisfactory, although lithology is undisputed.

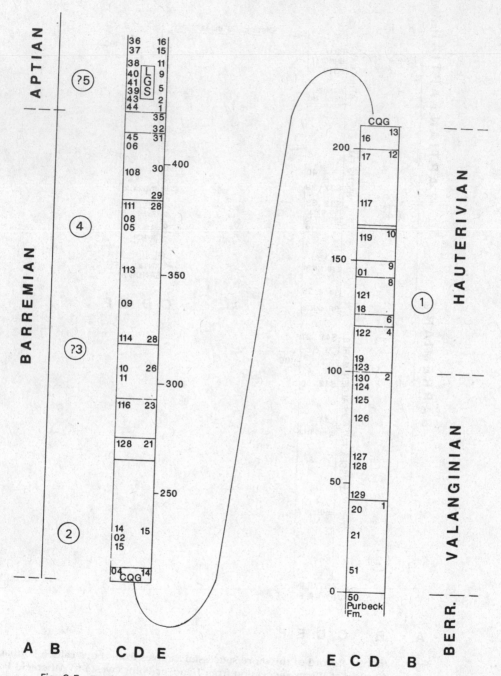

Fig. 9.5. Record of palynologic samples from the Wealden Formation of Worbarrow Bay, Dorset. A, ages of the geologic time-scale (GTS); B, phases 1–5 of the MCT assemblages; C, sample numbers; D, bed numbers (Arkell 1947); E, approximate rock thickness in metres above the top of the Purbeck formation. CQG, Coarse Quartz Grit (Bed 14); LGS, Lower Greensand.

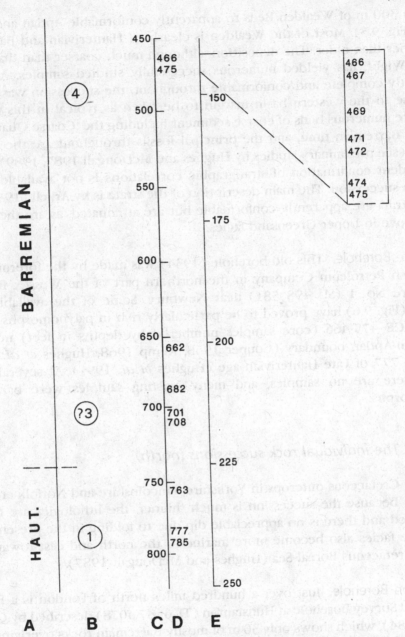

Fig. 9.6. Record of palynologic samples from Wealden strata of part of the Kingsclere Borehole. A, ages of the geologic time-scale (GTS); B, phases of MCT assemblages where evidence is available; C, borehole depth in feet; D, sample numbers; E, approximate borehole depths in metres.

through 300 m of Wealden Beds to apparently conformable Aptian and Albian strata (Fig. 9.5). Most of the Wealden is clearly of Hauterivian and Barremian age, rather than older. The deposition, although much coarser than that of the Isle of Wight, has yielded numerous successfully studied samples. Although apparently complete and conformable throughout, the succession was formed too close to the western basin margin to be taken as typical in this respect. There are numerous beds of coarse sediment including the 'Coarse Quartz Grit' in Early Barremian time, and the principal fossils throughout are the palyno- morphs (see preliminary studies by Hughes and McDougall 1987, 1989) so that independent confirmation of stratigraphic correlations is not available in this Wealden succession. The main description of the strata is by Arkell (1947); the Aptian strata are apparently conformable but are attenuated, as are the Albian strata above in Upper Greensand facies.

Kingsclere Borehole. This old borehole (1934) was made by the forerunner of the British Petroleum Company in the northern part of the Wessex Basin at Kingsclere No. 1 (SU 498 581) near Newbury. Some of the available core samples (Fig. 9.6) have proved to be particularly rich in palynomorphs at two levels: KCE 475-466 (core samples numbered by depths in feet) near the Barremian/Aptian boundary (Couper 1958, Kemp 1968, Hughes *et al.* 1979) and KCE 777 of Late Hauterivian age (Hughes *et al.* 1991). At several other levels there are no samples, and many existing samples were barren of palynomorphs.

9.4   The individual rock successions (north)

The Early Cretaceous outcrops in Yorkshire, Lincolnshire and Norfolk are very restricted because the succession is much thinner, the lithologies are mostly fine grained and there is no appreciable dip due to folding in the present land areas. The facies also become more marine to the north and east toward the original Cretaceous Boreal Sea (Hughes and McDougall 1987).

Hunstanton Borehole. Just over a hundred miles north of London is a British Geological Survey Borehole at Hunstanton (TF 6587 4078) described by Gallois (1972, 1984), which shows only 30 m of mostly Barremian rocks overlying Late Hauterivian (Fig. 9.7) near the southwestern margin of the Cretaceous Eastern or North Sea Basin; a small number of significantly successful samples were recorded. Aptian rocks above consist of a relatively thin sandy bed (Carstone) without palynomorphs. Ten or fifteen miles south-east of Hunstanton are two even shorter similar successions in further boreholes at Gayton (TF 7280 1974) and nearby Marham to the south (Gallois 1971).

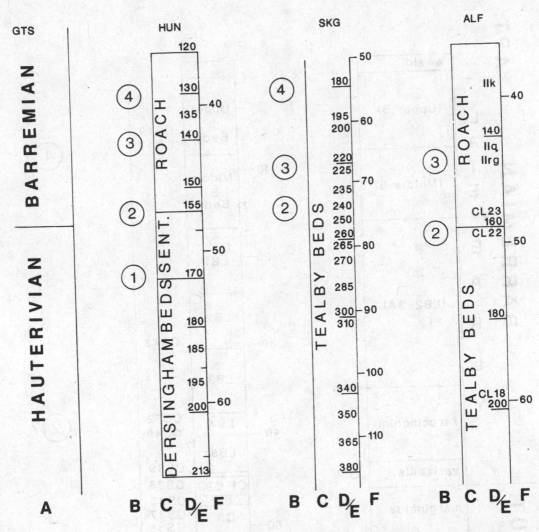

Fig. 9.7. Record of palynologic samples from relevant Early Cretaceous rocks of Hunstanton, Skegness and Alford Boreholes. A, ages of the geologic time-scale (GTS); B, Phases 1–4 of the MCT assemblages; C, rock formations; D, sample numbers; E, depths in feet; F, depths in metres. SNET?, ? Snettisham Clay. Scale differs in Skegness Borehole.

Skegness Borehole. Across the Wash on the Lincolnshire coast, the British Geological Survey Borehole at Skegness (TF 5711 6398) showed early Barremian and Hauterivian material (Gallois 1972), but from cuttings only.

Alford Borehole. This early (pre-1939) borehole near Alford (TF 460760) is about ten miles northwest of Skegness. Core samples of Valanginian, Hauterivian and Barremian age were at many levels marine in a glauconite facies with a full dinocyst presence. Pollen was distinctly less common than at Hunstanton, but the increased dinocyst record provided better stratigraphic correlation.

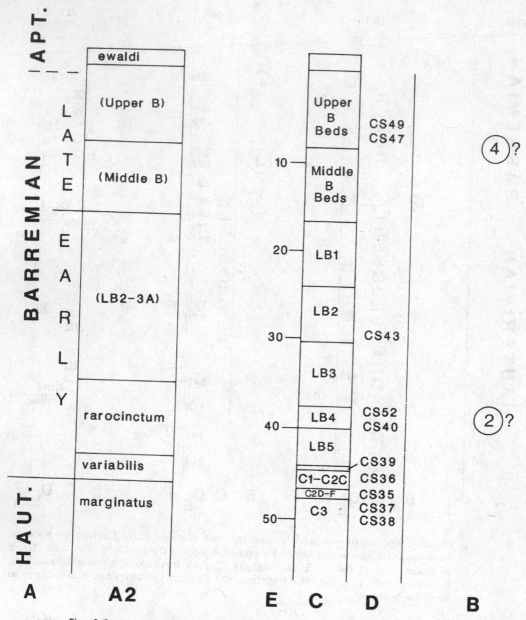

Fig. 9.8. Record of palynologic samples from Hauterivian–Barremian rocks of the marine Speeton succession (East Yorkshire coast); details taken from Neale (1974) and Harding (1990). A, ages of the geologic time-scale (GTS); A2, ammonite zones; B, approximate level of phases of the MCT assemblages; C, rock formation description; D, sample numbers; E, depths in metres below 'ewaldi marl'.

Speeton. Just south of Filey in East Yorkshire, the Speeton coastal section (TA 143764) was sampled from Hauterivian and Early Barremian strata (Fig. 9.8). This well-known succession is clearly documented (Neale 1974) and contains abundant marine megafossils and also dinocysts; correspondingly the pollen occurrences are sparser, resulting in much less scope for comprehensive interpretation. This succession is in places both faulted and severely landslipped, but is a unique outcrop representative of the western edge of a large basin now covered by the North Sea.

9.5   Geologic assessment

The samples from these successions have been assembled and worked over the last thirty-five years, first by light microscope reconnaissance and subsequently in more detail. There are clearly some other possibilities of samples like the Weald Clay brickpits of the Northern Weald and certain additional boreholes that in some cases have already proved disappointing in fossil content or too discontinuous and difficult to correlate. The described set (see Fig. 9.2) has been regarded as adequate for the present level of study of the Hauterivian to Aptian occurrences of these angiospermid monosulcates.

9.6   Method of preparation

The palynomorphs are extracted from clastic sedimentary rock by a process that has changed a little but has been kept essentially the same here for over thirty years. The cleanly crushed rock sample is treated successively with cold dilute hydrochloric acid, with hydrofluoric acid, and then with hot stronger hydrochloric acid to disperse fluorides; these parts of the process are assumed not to have damaged or altered the organic matter significantly. After test slides of palynofacies have been examined, oxidation of the extracted organic matter is made with concentrated nitric acid (or occasionally *in extremis* with other oxidants as necessary); this main critical step is kept to an absolute minimum duration consistent with freeing organic particles from one another. Usually one or two minutes treatment is employed, sometimes less (or rarely more) if test slides have shown the need; preparations made as far back as 1954 by Couper (1958) and others employed much longer times to gain translucency of the palynomorphs, with consequent heavy swelling and damage (recognised with hindsight), but these have all been subsequently reprepared from the samples. Very dilute ammonia solution is used very briefly to disperse oxidation products and to restore pH; this is in preference to potassium hydroxide for the same purpose, which is suspected of causing swelling of exines; the ammonia step is not regarded as critical, although it may appear to be so if unnecessary damage has already been caused by excessive oxidation. Centrifuging between washes is used in preference to slower methods, but obviously sieving is avoided with such small pollen under search. Zinc bromide is used as a heavy liquid when necessary to separate surviving non-silicate mineral grains. Light microscope slides are always made using Hydramount and Depex; earlier Clearcol was employed, and before that glycerine jelly with a coverslip sealant. Residues are stored in glycerine/water with a very small amount of phenol for protection.

9.7   *Microscope examination*

Light microscope slides are examined for general palynofacies data, but direct use of palynofacies in either selection or rejection of preparations for search has not always been found helpful; the find recorded by Hughes *et al.* (1991) came for example from an unusual preparation therein described. Discrimination of this very small pollen (<20 μm) with light microscopy has proved to be unsatisfactory and is no longer attempted: Couper (1958) found only a handful of grains in several samples, despite his skill as an observer; Kemp (1968) by great perseverance found a hundred specimens of Couper's *Clavatipollenites hughesii* after examining twenty slides from one sample. Instead, strew-mounts are now made on SEM stubs which have already been fitted (Fig. 9.9) with Cambridge nickel grids (Mark 3); direct strew-search is carried out by SEM with photographic record, grid co-ordinates for each specimen being read by the observer. Some palynomorphs have additionally been studied by TEM, but as the main thrust of this operation is essentially for continuous SEM search and recordings (to replace light microscopy), no attempt has been made to emulate the elegant single grain studies of Walker and Walker (1984, 1986). The recorded fossil material consisting of palynomorphs mounted on gridded stubs and coated, appears to endure satisfactorily. Some material that has already been held for seventeen years after original photography can still be used normally for reobservation with recoating not always necessary. Admittedly some light microscope slides are thirty-five years old and still usable, but failures are fairly common because processing was rougher and less well understood in the fifties. Palynomorph records have to rely in practice on photographs and other illustrations, and there seems to be no reason why properly stored stubs and negatives should be any more hazardous to preserve as back-up than chemical mountants on glass slides.

In SEM strew-search observing, the required monosulcate and other pollen grains are much smaller (10–20 μm) than the normal background of standard spore, pollen, dinocyst, wood and cuticle fragments with average unit size of 50 μm or more. This obliges the observer to select carefully a working magnification for which the stub-grid pattern has also been attuned. To keep observation time at an economic level, selection of types of palynomorph for study and

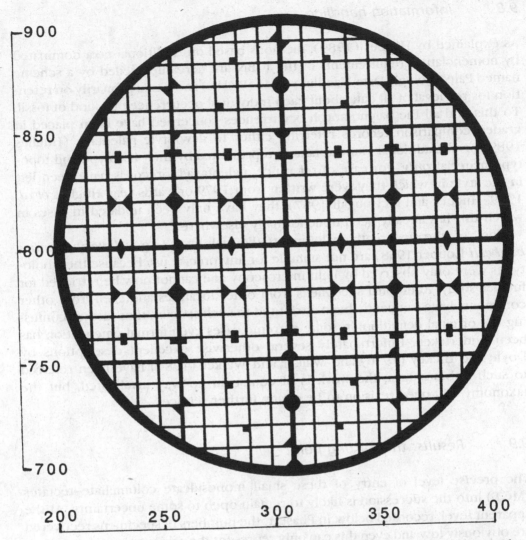

Fig. 9.9. Enlarged reproduction of specially marked nickel stub-grid (Cambridge, Mark 3), which is used to enable the observer to undertake an organised strew-search and to locate selected specimens for study and photography. The co-ordinate scale in 'eastings' and 'northings' is drawn on the record pad, and fits the special marks, which can easily be found.

photography is carefully restricted, although time must also be allowed for manipulation such as stub-tilting to attempt to record apertures that may otherwise not be ideally displayed. Information lines and scale marks of the SEM photographs are retained permanently on prints for record and are deliberately used in publication.

9.8   Information handling

As explained by Hughes (1989), methods based on traditional taxa dominated by nomenclature requirements of the ICBN are here superseded by a scheme named Paleontologic Data-Handling Code (PDHC) and based primarily on retention for retrieval of all data about each individual occurrence of a kind of fossil. To this end all the palynomorph occurrences concerned have been placed in graded comparison records referring either to new style paleotaxa (Hughes 1989) or to traditional Linnean taxa as represented by their holotype and topotype material *only*. For ease of reference, paleotaxa/biorecords have been left in the style in which they were written from 1979 onwards (e.g. Hughes *et al*. 1979, Hughes and McDougall 1987); they have only been updated in cases in which this has already been done formally elsewhere.

Most of the apparently relevant traditional taxa (e.g. *Clavatipollenites hughesii* Couper 1958) are not suitable for any precise use because their holotypes were only observed by light microscopy and cannot now be retrieved for further study. Additional specimens from other localities and even from other continents have frequently been attributed to such species, thus further confusing the original definition; in some extreme cases even formal emendation has been undertaken. Unfortunately several otherwise excellent descriptions by Doyle and Hickey (1976) and Walker and Walker (1984) have been referred to such dubious taxa; the work is of course in no way invalidated, but the taxonomy will need revision to be of any further service.

9.9   Results: the starting point

The precise level of entry of these small monosulcate columellate–tectates (MCT) into the succession is likely to remain open to some uncertainty. At the apparent level, recorded below in Phase 1, the numbers of specimens recovered are obviously low and even this can only represent the achievement of a particular (unknown) frequency of occurrence for the plants concerned. The selection of Phase 1 in Late Hauterivian time is simply a matter of judgement, although it should not be moved earlier (lower) unless new finds add up to a redefinition of Phase 1. Very rare earlier occurrences of occasional palynomorph specimens are to be expected, but they will not achieve significance unless they are newly recorded in more than two correlatable parts of successions. On the practical side it is important to recognise observer fatigue: efficiency of search clearly falls off when the results remain negative for long periods. There can also be uncertainty about the precise nature of objects being sought when the tectum could well be virtually entire and any columellae might be completely hidden. Stimulus for search is probably best attributed to other discoveries such as that of Trevisan (1988), who reported some simple Phase 1 MCT palynomorphs from dinocyst-dated Valanginian rocks in western Italy, although full documentation is expected shortly. The more dramatic discoveries in Late Triassic rocks (Cornet 1989*b*) or Late Jurassic rocks (Cornet and Habib 1992) were not of early MCT palynomorphs and are discussed elsewhere (Chapter 6).

There is no obvious change of sedimentary facies near the Phase 1 starting point, and persistent other fossils such as C- or S-phase ostracods or pterido-phyte palynomorphs do not indicate significant changes just below this level. Having established an MCT pattern of occurrence, described below, the greater priority is now to seek out evidence for the MCT plant or plants, although more palynomorph discoveries such as that of Hughes *et al.* (1991) clearly help.

9.10   Results: the phase presentation

The successive four phases of MCT development, plus one phase below and one above for delimitation, are each defined in words in the text and are each referred to a key sample and preparation to show adequate presence of the palynomorphs concerned and adequate quality of preservation. The earlier of the key samples are from the Warlingham Borehole, south of London, which was initially selected as reference because the rock succession appeared to be continuous from Late Jurassic through to Cretaceous Aptian rocks. The other successions extend and amplify the Warlingham record, and in later phases replace it. To maintain stratigraphic order, Phase 0 is described first as an attempt to record possible links between unsculptured monosulcates and the MCT monosulcates under study, or more simply to assess the mid-Hauterivian base assemblage from which the MCT in Phase 1 evolved. The phases in use, with brief definitions, are in the table below.

Age	Phase	Title	Definition
APT	Phase 5	Consolidation	Angiospermid pollen up to 10% of count
BARR	Phase 4	MCT 4 climax	Twelve taxa diversity
	Phase 3	MCT 3	Crotonoids enter, and six taxa diversity
	Phase 2	MCT 2	Numerous specimens available to erect more than one taxon
HAUT	Phase 1	MCT 1 start	Record of individual specimens in more than two successions
	Phase 0	Search below	Includes occurrences of lesser extent than Phase 1 definition

APT, Aptian; BARR, Barremian; HAUT, Hauterivian.

It is not intended that the phases of development should be regarded in any way as zones or be used as such. No attempt therefore has been made to demar-cate boundaries between phases; any such boundaries needed subsequently should be constructed by including the ranges of other palynomorphs present and combining the results. This preserves the freedom to characterise the phases independently, and underlines the idea that stratigraphic subdivisions should be sought only after amalgamation of all available data and not credited afresh for each speciality as it arises. Separate erection of palynevents for stratigraphic correlation (see Hughes 1989) is not precluded as the representatives and mentioned samples for phases are the best so far discovered.

9.11   Succession of phase detail: Hauterivian

Phase 0: search below. Normally there are no MCT specimens recorded and no *Ephedripites*; if any such specimens are found, the sample/preparation remains in Phase 0 unless the requirement for Phase 1 is met. The favourable successions for study of the nature of pre-MCT palynomorphs and any possible intermediates are Worbarrow Bay, although much of the relevant deposition there is of a coarse clastic type, and the Warlingham Borehole, which includes at this level some minor faulting and thus probable repetitions and omissions of strata. The general content of larger spores and pollen of assemblages from Warlingham 2000 up to 1500 was studied by light microscope and published (Hughes and Moody-Stuart 1969, Batten 1973, Hughes and Croxton 1973). *Dicheiropollis*, first described in pre-Barremian strata from Italy (see Trevisan 1971) and then from North and West Africa, has not yet been encountered in England. The present SEM strew-search study was concentrated on small pollen of less than 25 µm diameter which has sculptured exine and has not previously been described; several types were recorded by Hughes and McDougall (1987). There is no evidence to suggest any possible plant group affinity, although some of them may prove to be tectate in a simple way. Figs. 9.10, 9.11 and 9.12 illustrate a group of such fossils from the samples WM1524/9 and WM1512/8 including CfA Hauterivian-*microtect*, CfB Hauterivian-*cactisulc*, and Hauterivian-*lacebee*, all of which appear to be monosulcate–tectate and none of which exceed 18 µm maximum diameter. Figure 9.12 at lower magnification includes several other specimens to indicate some of the range of variety displayed in these fossils that occur commonly in most samples of this phase and therefore probably originated much earlier. It will probably not be worth speculating on their provenance until their full range is more accurately known. Occasional isolated columellate grains, which would be acceptable in MCT, do occur and have been recorded (e.g. Fig. 9.11, 0479); they may indicate some convergence with the views of Trevisan (1988, 1994), mentioned above. Nothing comparable with the diporate pollen of Taylor *et al.* (1987) has yet been seen.

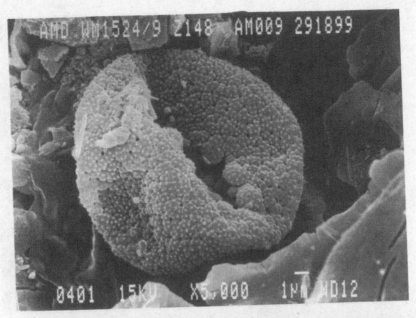

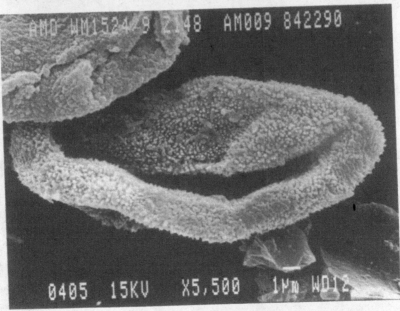

Fig. 9.10. Phase 0, scanning electron micrograph (SEMG) of Monosulc group, Warlingham Borehole: CfB Hauterivian-*cactisulc* (0401); Paleotaxon Hauterivian-*dimorph* (0405). Notes on photographs: observer identification, sample number, preparation number, stub number, stub-grid co-ordinates. Bottom line: photoserial number (Jeol JSM820), KV, original magnification, scale bar value, reference number. Other data are filed in the Department of Earth Sciences, University of Cambridge.

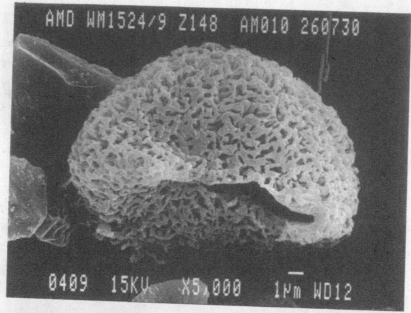

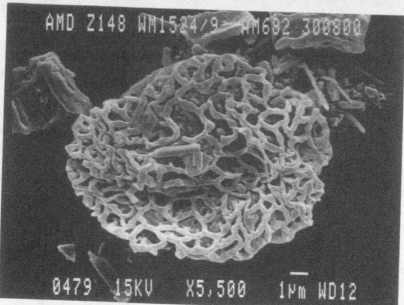

Fig. 9.11. **Phase 0. SEMG of Monosulc group, Warlingham Borehole: Paleotaxon Hauterivian-*lacebee* (0409); CfB Retisulc-*newling* (0479).** For notes, see explanation of Fig. 9.10.

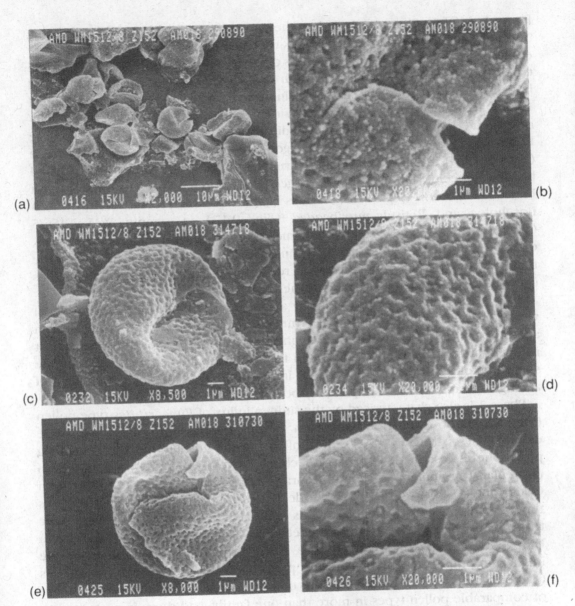

Fig. 9.12. Phase 0, SEMG of Monosulc group, Warlingham Borehole, Paleotaxon Hauterivian-*microtect*: (a) assemblage (0416); (b) aperture detail (0418); (c) specimen (0232) and (d) detail (0234); (e) specimen (0425) and (f) detail (0426). For notes, see explanation of Fig. 9.10.

Phase 1: MCT 1. Phase 1 is based on the same or comparable MCT fossils being observed in more than two successions representing Late Hauterivian time. The earliest discovered MCT palynomorphs disconcertingly proved to be of unusual types (not found in later phases) with coarsely banded or beaded muri, and initially they were found as scattered single specimens (Hughes and McDougall 1987), so that their significance remained doubtful. Patience was eventually rewarded (Hughes *et al.* 1991) by the discovery in sample KCE777 (Kingsclere Borehole) of an adequate set of ten good specimens of Hauterivian-*colthick* (Figs. 9.13, 9.14), which provided firm support for the previously isolated banded specimens and for a general correlation. The paleotaxon Retisulc-*muriverm* (Figs. 9.15, 9.16) is based in sample HUN170 (Hunstanton Borehole) and Retisulc-*muribeaded* (Fig. 9.17) in sample WM 1456/5 (Warlingham Borehole). Isolated specimens compared with standard later types, e.g. CfA Retisulc-*newling* (Fig. 9.18) and CfB Retisulc-*laevgat* accompany the more elaborate banded-muri taxa but remain at this level as single specimens even after careful examination of further SEM stubs.

This recorded variety of supposedly angiospermid pollen, although in detail weakly based, is important not so much in indicating origin from unseen ancestors that might have existed slightly earlier, but in deflating the almost routine assumptions that they have all to be considered as radiating from one source taxon. This theorising is based in the dogma of monophyletic origin of all angiosperms, which as explained above (Chapter 8), is here discarded.

Phase 1 is also characterised by the sudden influx of *Ephedripites*-type pollen (Fig. 9.19), of completely unknown origin, despite the suggestive name employed (see Hughes 1992); the pollen persists through all the remaining phases here described, in what are clearly several separate taxa. It is also particularly common in South Laurasia (Trevisan 1980) and Northern Gondwana (Osborn *et al.* 1993), but decreases everywhere in Late Cretaceous time. Palynologists have recorded this general type of pollen (not the same taxa) from Triassic time, from mid-Cretaceous time and from Late Tertiary to Holocene *Ephedra* and *Welwitschia*, with little or no record from the intervening times such as the Jurassic period and no pre-Pliocene megafossils (Harland *et al.* 1967). The angiosperms offer many cases of parallel development of comparable pollen types in more than one family, and there is no reason to expect less variety and diversity from Mesozoic Pollenifera. These *Ephedripites* pollen of Cretaceous age are best treated as another interesting group of unknown origin. Finally there are comparison records of some of the earlier taxa, e.g. CfA Hauterivian-*cactisulc* (Fig. 9.20).

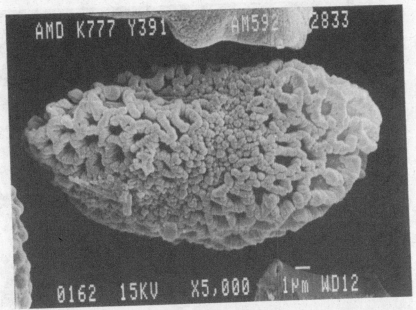

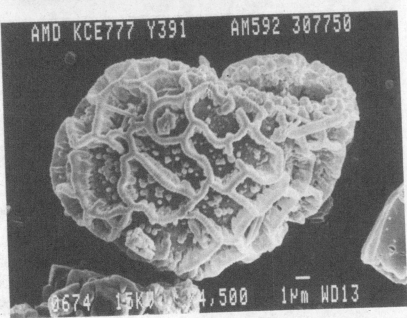

Fig. 9.13. Phase 1. SEMG of Retisulc group, Kingsclere Borehole: Paleotaxon Hauterivian-*colthick*; aperture (0162), non-apertural detail (0674). For notes see explanation of Fig. 9.10.

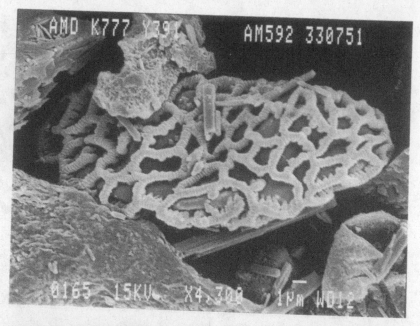

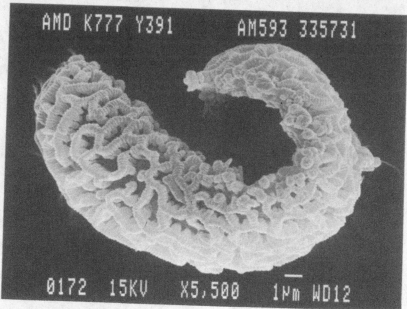

Fig. 9.14. Phase 1. SEMG of Retisulc group, Kingsclere Borehole: Paleotaxon Hauterivian-*colthick*; columellae (0165), open aperture (0172). For notes, see explanation of Fig. 9.10.

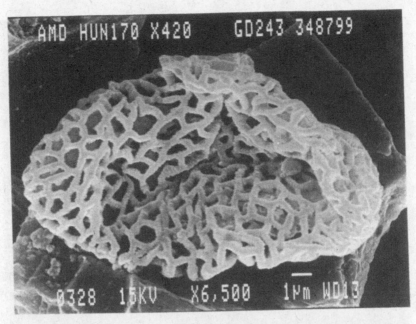

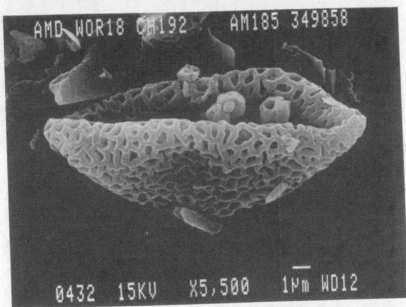

Fig. 9.15. Phase 1. SEMG of Retisulc group: Paleotaxon Retisulc-*muriverm* (0328), Hunstanton Borehole; CfA Retisulc-*muriverm* (0432), Worbarrow. For notes, see explanation of Fig. 9.10.

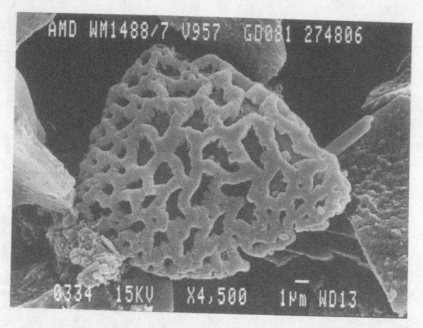

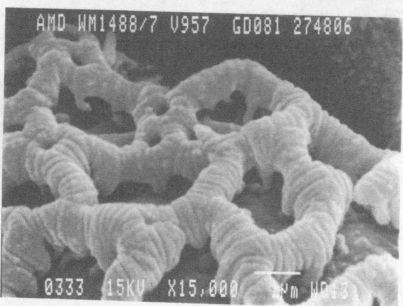

Fig. 9.16. Phase 1. SEMG of Retisulc group, Warlingham Borehole: CfA Retisulc-*muriverm* (0334); detail of same (0333). For notes, see explanation of Fig. 9.10.

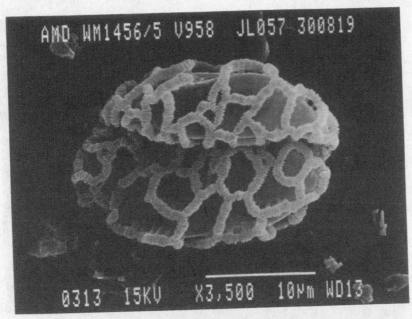

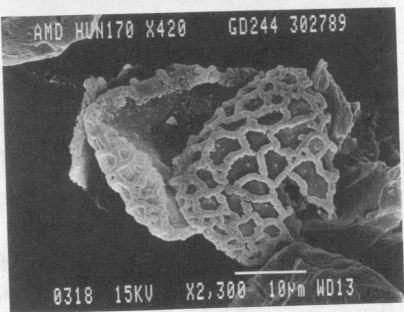

Fig. 9.17. Phase 1. SEMG of Retisulc group: paleotaxon Retisulc-*muribeaded* (0313),
Warlingham Borehole; CfA Retisulc-*muribeaded* (0318), Hunstanton Borehole. For
notes, see explanation of Fig. 9.10.

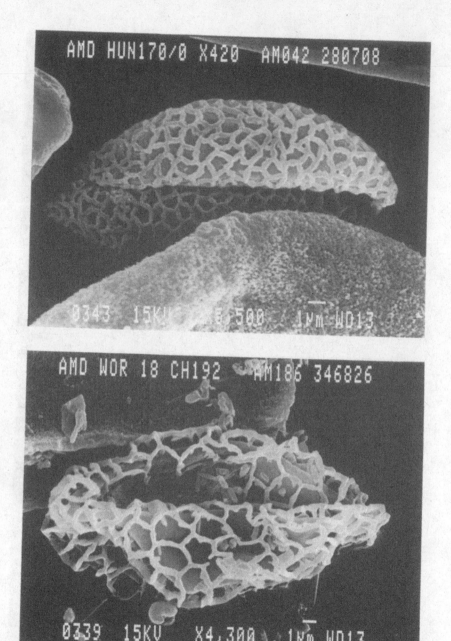

Fig. 9.18. Phase 1. SEMG of Retisulc group: CfA Retisulc-*newling*; from HUN170 (0343) and from WOR18 (0339). For notes, see explanation of Fig. 9.10.

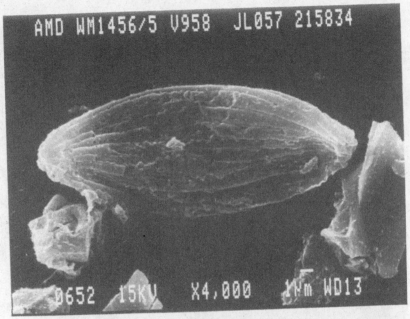

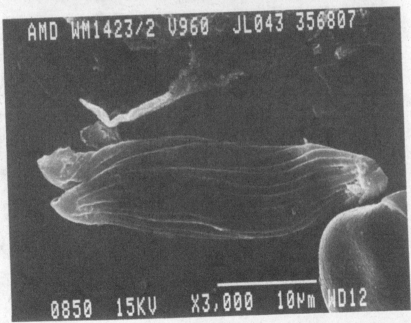

Fig. 9.19. Phase 1. SEMG of Ephedripites group, Warlingham Borehole: CfA Hauterivian-*pointboat* (0652); CfA Hauterivian-*pointboat* (0850). For notes, see explanation of Fig. 9.10.

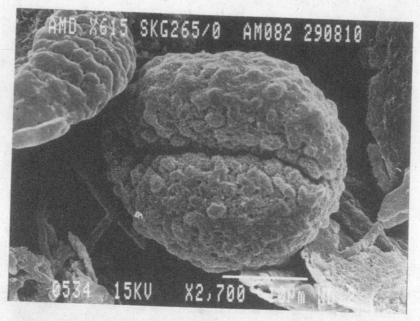

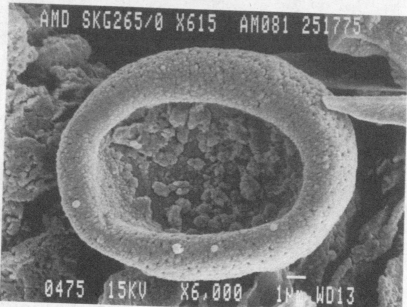

Fig. 9.20. Phase 1. SEMG of Monosulc group, Skegness Borehole: CfA Hauterivian-*cactisulc* (0534), SEMG of Ring Group: CfA Barremian-*ring* (0475). Notes on photographs: observer identification, sample number, preparation number, stub number, stub-grid co-ordinates. Bottom line: photoserial number (Jeol JSM820), KV, original magnification, scale bar value, reference number. Other data are filed in the Department of Earth Sciences, University of Cambridge.

9.12   Succession of phase detail: Barremian

Phase 2: MCT 2. This phase records for the first time the appearance in certain individual samples of (a) adequate numbers of MCT specimens to erect useful and convincing taxa, and (b) some regularity of occurrence of CfA comparison records through all relevant successions in the whole area studied. This suggests that the plants represented by these monosulcate columellate-tectates (MCT) were by this time present in most florules, which was obviously not the case in the previous Phase 1. The variety of MCT palynomorph taxa present was in general confined to comparison records of the three taxa Retisulc-*dentat*, Retisulc-*laevgat*, and Retisulc-*newling* (Hughes *et al.* 1979); all three possess lumina of about 1 μm diameter and thus tend to appear similar at low magnifications. They differ principally in the sculpture of the muri and to some extent in the height of the columellae (Hughes *et al.* 1979, p. 520).

The reference sample is from the Warlingham Borehole (WM 1415/6), and several adjacent samples from WM 1423/2 up to WM 1394/1 are similar although less rich. The commonest and most prominent pollen is Paleotaxon Retisulc-*dentat* founded on fifty specimens in a preparation of sample WM 1415/6 (Fig. 9.21). In the main 'Atherfield' succession of the Isle of Wight, the favourable sample S44 contains clusters of CfA Retisulc-dentat (Fig. 9.22), although unfortunately they do not reveal the form of the presumed anther; all samples from approximately the lowest exposed 100 m of this succession as represented by Beds 1–15 (White 1921) are within Phase 2. West of Hanover Point in the Lower Compton Bay succession, although Beds 1–12 there (of White 1921) are equivalent to the lower beds of the Atherfield succession, the lithologies are less favourable and there happen to be few successful samples. At Worbarrow in Dorset, the phase is displayed in samples WOR 004 and WOR 015 from Beds 14 and 15 (Arkell 1947); it is not strongly developed but is in any case closely above the 'Coarse Quartz Grit' in an interval of obviously very rapid deposition near the western edge of the Wessex Basin. Interestingly, the very unusual spore *Tesselatosporis escheri* described by Harding (1988) has also been recorded from Worbarrow Beds 8–12 (Arkell 1947) and may indicate (Fig. 9.23) Phase 2 representation through from Beds 8–15, although the duration of this actual deposition may have been short. In the area of the north-eastern basin, the Hunstanton Borehole samples HUN 160 and HUN 155 may be taken to represent Phase 2 with comparison records CfA Retisulc-*dentat* and CfA Retisulc-*newling* only. Further north in Skegness, Alford and Speeton, where dinocysts are universal and pollen yields correspondingly lower, the evidence allows for development of Phase 2 but does not positively confirm its presence. This applied also to Phase 1 in these successions and further search will be necessary to support this hypothesis. Rare specimens of other pollen sculptures occur (Fig. 9.24) and *Ephedripites* group taxa (Fig. 9.25) begin to diversify.

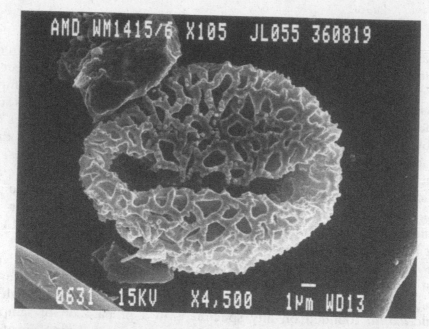

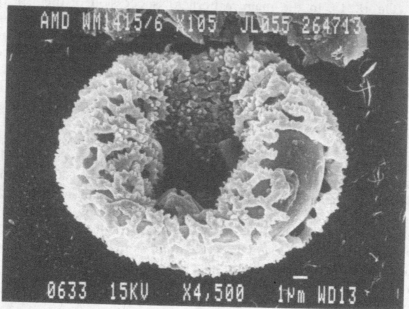

Fig. 9.21. Phase 2. SEMG of Retisulc group, Warlingham Borehole: Paleotaxon Retisulc-*dentat* (0631 and 0633). For notes, see explanation of Fig. 9.20.

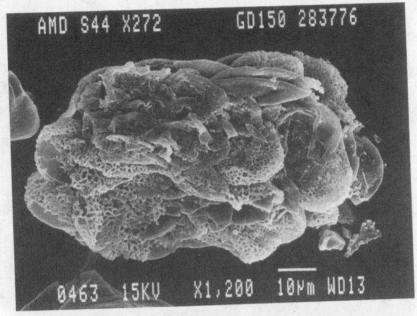

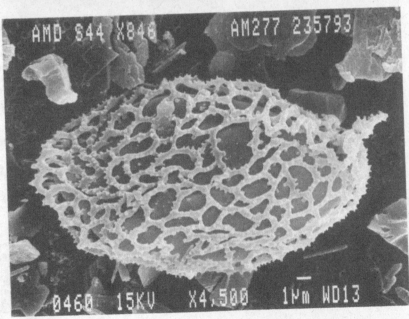

Fig. 9.22.   Phase 2. SEMG of Retisulc group, Atherfield succession, Isle of Wight: CfA Retisulc-*dentat*, group (0463) and single specimen (0460). For notes, see explanation of Fig. 9.20.

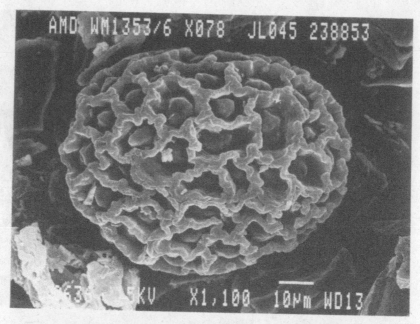

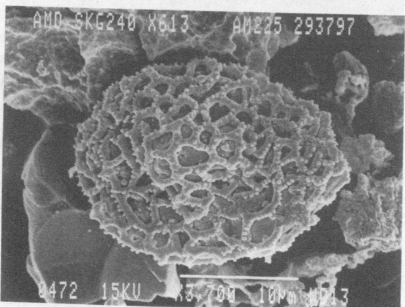

Fig. 9.23. Phase 2. SEMG of CfA *Tesselatosporis escheri* (0636), Warlingham Borehole. SEMG of CfB Retisulc-*dentat* (0472), Skegness Borehole. For notes, see explanation of Fig. 9.20.

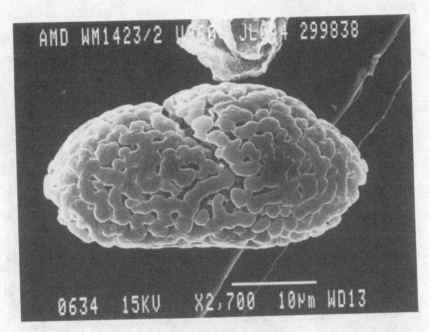

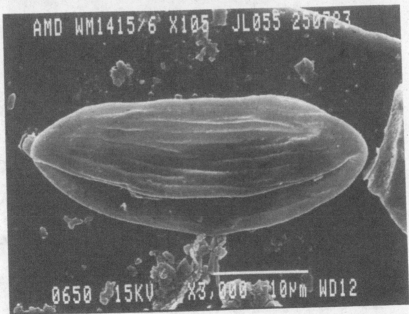

Fig. 9.24. Phase 2. SEMG of Retisulc group, Warlingham Borehole: CfB Retisulc-*laevgat* (0634). SEMG of Ephedripites group, Warlingham Borehole; CfA Barremian-*monosixteen* (0650). For notes, see explanation of Fig. 9.20.

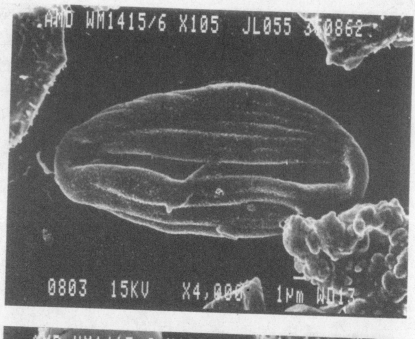

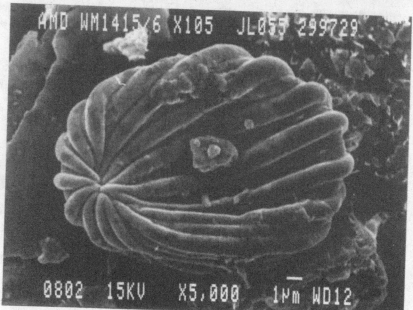

Fig. 9.25. Phase 2. SEMG of Ephedripites group, Warlingham Borehole; Paleotaxon Barremian-*monosixteen* (0803); CfA Barremian-*canalic* (0802). For notes, see explanation of Fig. 9.20.

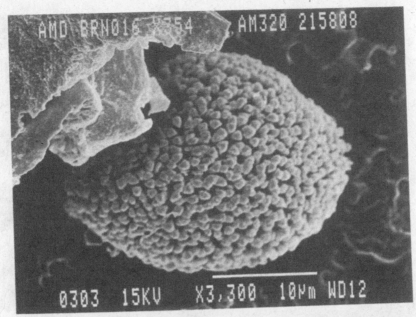

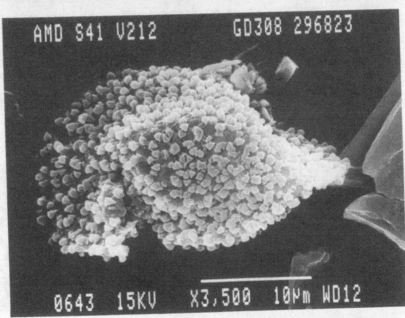

Fig. 9.26. Phase 3. SEMG of Superret group, Atherfield succession, Isle of Wight: CfA Superret-*croton* (0303); CfA Superret-*croton* (0643). For notes, see explanation of Fig. 9.20.

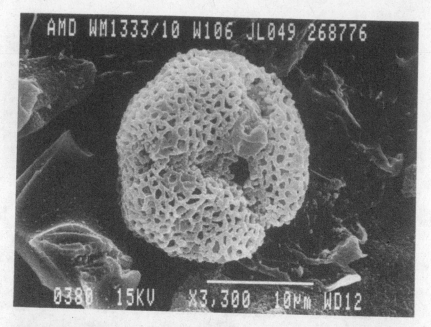

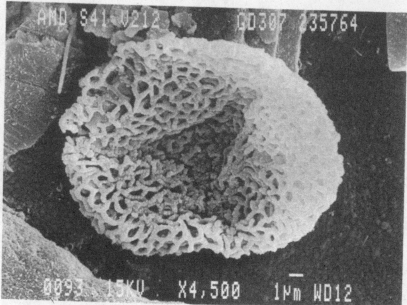

Fig. 9.27. Phase 3. SEMG of Retichot group: CfA Retichot-*baccat* (0380), Warlingham Borehole; CfA Retichot-*baccat* (0093), Atherfield succession, Isle of Wight. For notes, see explanation of Fig. 9.20.

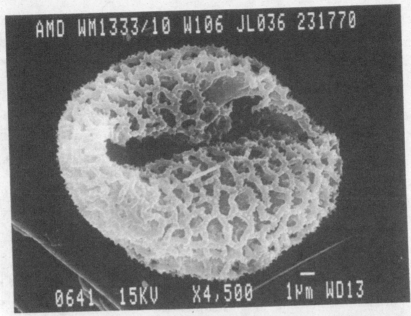

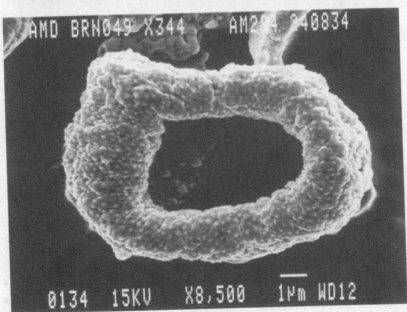

Fig. 9.28. Phase 3. SEMG of Retisulc group: CfA Retisulc-*dentat* (0641), Warlingham Borehole. SEMG of Ring group: CfA Barremian-*ring* (0134), Atherfield succession, Isle of Wight. For notes, see explanation of Fig. 9.20.

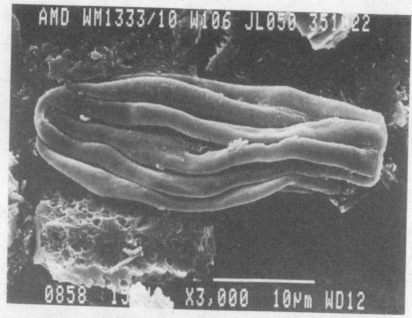

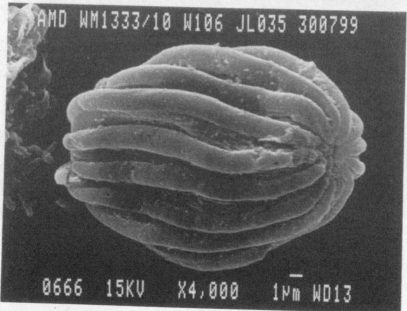

Fig. 9.29. Phase 3. SEMG of Ephedripites group, Warlingham Borehole: CfA Barremian-*monosix* (0858); CfA Barremian-*canalic* (0666). For notes, see explanation of Fig. 9.20.

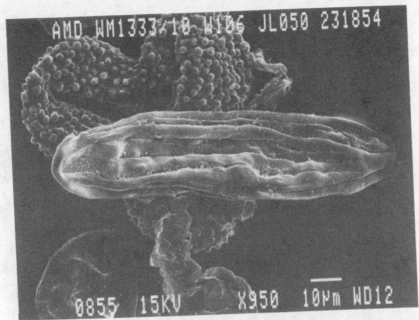

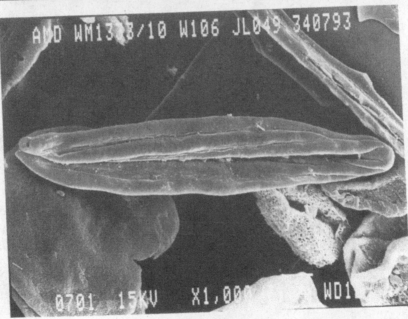

Fig. 9.30. Phase 3. SEMG of Ephedripites group, Warlingham Borehole: Paleotaxon Barremian-*megaspindle* (0855 and 0701). Notes on photographs: observer identification, sample number, preparation number, stub number, stub-grid co-ordinates. Bottom line: photoserial number (Jeol JSM820), KV, original magnification, scale bar value, reference number. Other data are filed in the Department of Earth Sciences, University of Cambridge.

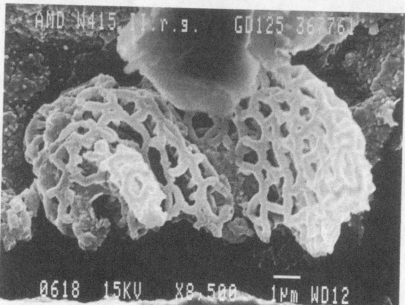

Fig. 9.31. Phase 3. SEMG of Retisulc group: CfA Barremian-*teebac* (0104), Atherfield succession, CfA Barremian-*teebac* (0618), Alford Borehole, 146 feet. For notes, see explanation of Fig. 9.30.

Phase 3: MCT 3. This phase includes records of the earliest assemblages, which show a regular diversity of from six to eight distinct pollen types; they include (Fig. 9.26) the first crotonoids (CfA Superret-*croton*) and the first forms that frequently display a trichotomosulcate rather than a monosulcate aperture. The latter have been recorded (Hughes *et al.* 1979) under Paleotaxon Retichot-*baccat* (Fig. 9.27). A further new form is Barremian-*ring* (Fig. 9.28), which could perhaps also be fairly recorded as CfB *Tucanopollis crisopolensis*; this species is already known from rocks of this general age in West Africa (Doyle *et al.* 1977) and in Brazil (Regali *et al.* 1974). Comparison records (CfA) of the three taxa new in Phase 2 remain throughout Phase 3, which is, however, noticeably richer. The reference sample is conveniently again taken from the Warlingham Borehole (WM 1333/10), and is rich in records of *Ephedripites* group taxa (Fig. 9.29) including the unusually elongated Barremian-*megaspindle* (Fig. 9.30), which is up to 90 μm long. From the main 'Atherfield' succession of the southern Isle of Wight, the sample S41 (see Figs. 9.1, 9.4), from Bed 21 in the upper part of the Wessex Formation (Wealden marls) provides a typical Phase 3 assemblage (Fig. 9.31); other samples from Beds 21 to 32 are comparable. The assemblage from sample LCB16 of the Lower Compton Bay succession west of Hanover Point (Bed 13) also represents this phase. North-west of the fault the Upper Compton Bay succession has probably not yielded assemblages of this phase because landslipping in the coastal section has prevented organised sample collection from the relevant beds. The northern boreholes of Hunstanton and Alford (Fig. 9.31) have provided recognisable representation of this phase, although in a form diluted by marine influence.

9.13   Succession of phase detail: Barremian–Aptian

Phase 4: MCT 4. The assemblages indicating Phase 4 show the culmination of Barremian diversity in the MCT pollen group, and are present at about the Barremian/Aptian boundary in almost all the English successions studied. There are normally up to twelve distinct pollen types and indications of others yet to be described. Although it would still be possible to draw a reference sample from the Warlingham Borehole (WM 1105/4–8), the richest samples are undoubtedly from the Kingsclere Borehole (KCE 475 to KCE 466); unfortunately the rock samples from immediately below and above this short section of core have never been available so that no stratigraphic limits can be provided. It is therefore more suitable to use the Isle of Wight successions for reference, in particular sample S38 from Bed 35 of the Vectis Formation (Wealden Shales) of the main Atherfield succession in which Phases 2–5 inclusive are encountered in a continuously exposed low-dip coastal outcrop. There are also good supporting samples in the other Isle of Wight successions: samples S47 and UCB43 of the Upper Compton Wealden, which is fault-bounded below; samples LCB12 and LCB6 of the Lower Compton Wealden, which is fault-bounded above; and samples SYB27 and SYB24 of the Sandown Wealden, which is anticline-bounded close below.

Interestingly, in the Atherfield and Sandown successions, Kerth and Hailwood (1988) on paleomagnetic evidence have claimed that the age of much of the Vectis Formation is earliest Aptian (= fissicostatus-zone but without any ammonites) rather than the traditional Barremian age and this would include sample S38 in rocks of Aptian date; unfortunately the magnetic study has not yet been applied to the Compton Bay successions. Phase 4 would as a result straddle the Barremian/Aptian boundary, and the Perna Bed would be well within the Aptian age, affecting stratigraphic interpretations further afield such as that in the Warlingham Borehole; such interpretations remain to be tested on general grounds but the paleobotanical history is unaffected except as to stratigraphic scale calibration.

The Phase 4 climax assemblage composition of MCT taxa may be grouped for general discussion as follows: (a) Retisulc group forms from Phases 2 and 3 plus apparent derivations; (b) Retisulc group new forms: crochet, perfotect and others; (c) Crotonoid group; (d) miscellaneous.

(a)   Comparison records (CfA) of Retisulc-*dentat*, Retisulc-*laevgat*, Retis-ulc-*newling* and Retichot-*baccat* from Phases 2 and 3 are present but not prominent in Phase 4. The simple derivatives are Paleotaxon Aptian-*dentsharp* (Figs. 9.32, 9.33) from Hunstanton Borehole (HUN130) and a CfB form from the Alford Borehole (146/IIrg), with a denser development of longer processes on the muri than in Retis-ulc-*dentat*. Paleotaxon Retisulc-*monbac* was originally described (Hughes *et al*. 1979) from Kingsclere Borehole (KCE 474 and 475), and two Isle of Wight CfA records are provided here (Fig. 9.34). Finally from Warlingham Borehole (WM 1105/4–8) is a record CfB Retichot-*baccat* (Fig. 9.35) with strikingly blunt mural processes.

(b)   Not resembling any earlier occurrences, are first Paleotaxon Retisulc-*crochet* from the Kingsclere Borehole (KCE 474), which displays two lumen sizes evenly distributed (Fig. 9.36), and CfB Retisulc-*crochet* from the Atherfield successions (S37) in which there is a solid aper-ture margin and some concentration of small lumina adjacent to this. Quite distinct (Fig. 9.37) is CfC Retisulc-*crochet* from Kingsclere (KCE 474), with very broad muri, short columellae, and numerous aborted columellae in mid-lumen. In contrast (Fig. 9.37) Paleotaxon Aptian-*longcol* from Kingsclere Borehole (KCE 467) has narrow muri, very long columellae, and two lumen sizes with CfA records (Figs. 9.38, 9.39) from the Isle of Wight, from Upper Compton (S47) and Sandown (SYB27). Also new at this Phase 4 horizon are Paleotaxon Aptian-*perfotect* from Kingsclere Borehole (KCE 474), of which any earlier CfA records are now known to be in error. This is more nearly tectate than semi-tectate, with variable diameter perforations with

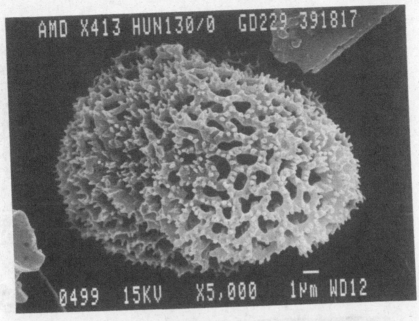

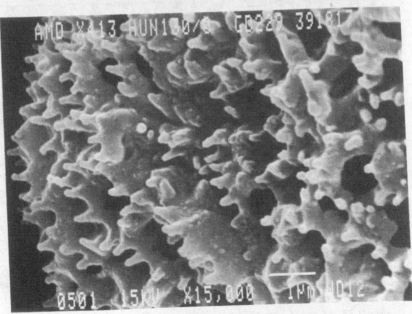

Fig. 9.32. Phase 4. SEMG of Retisulc group, Hunstanton Borehole: Paleotaxon Aptian-*dentsharp* (0499) and enlargement (0501). For notes, see explanation of Fig. 9.30.

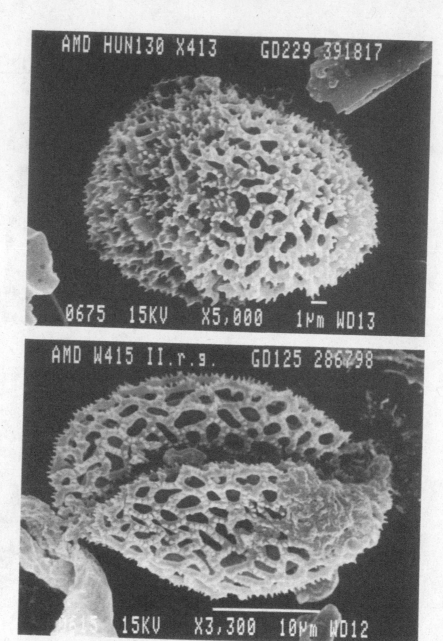

Fig. 9.33. Phase 4. SEMG of Retisulc group: Hunstanton Borehole, Paleotaxon Aptian-*dentsharp* (0675); Alford Borehole 146 feet II rg. CfA Aptian-*dentsharp* (0615). For notes, see explanation of Fig. 9.30.

monosulcate CfA examples (Figs. 9.40, 9.41) from Hunstanton Borehole (HUN130) and Sandown (SYB24) plus a trichotomosulcate form from Sandown (SYB31) similar to one from Kingsclere (KCE469) from Hughes *et al*. (1979, plate 64, figs. 7–8). A new paleotaxon, Aptian-*frill* (Fig. 9.42) from Isle of Wight Sandown (SYB24), displays a new style of laterally sculptured and therefore wide muri on short columellae.

(c)   Crotonoids were first observed and named (Hughes *et al*. 1979) from Kingsclere Borehole samples (KCE475, KCE474) from a succession that is now classed as Phase 4, before the discovery that CfA Superret-*croton* entered in Phase 3 (see Fig. 9.26). They may also have contributed to early light microscope observations on sample KCE475 by Couper (1958), from which he raised the much-used name *Clavatipollenites hughesii*; consequently there is uncertainty over the continued use of that name, even with qualifications devised by Doyle *et al*. (1977) for simpler pollen. CfA Superret-*croton* from Isle of Wight Atherfield (BRN147) is compared (Fig. 9.43) with a specimen from the paleotaxon (KCE475); other interesting developments (Fig. 9.44) are suprasculpture on top of the supramural crotonoid elements from Atherfield (BRN016), and flat muri devoid of the crotonoid elements from Isle of Wight Sandown (SYB31). The Paleotaxon Superret-*krinkel*, in which many of the crotonoid elements are fused together was erected by Hughes *et al*. (1979) on a Kingsclere sample (KCE 474), but the CfA record figured here is from the Worbarrow succession (WOR 038) and is included with Phase 5 described below.

Although some of these variants may fail as taxa from later observations, their Phase 4 occurrence in different successions suggests otherwise. Entirely distinct (Figs. 9.45, 9.46) is Paleotaxon Superret-*subcrot* from the Kingsclere Borehole (KCE474), in which the crotonoid elements are so small, numerous and closely packed that the taxon is difficult to recognise except at higher SEM magnification than it is convenient to use in scan search.

(d)   Miscellaneous additional records include a tetrad (Fig. 9.46) of CfA *Eucommiidites minor* confirming the orientation of this pollen, a spectacular monosulcate (Fig. 9.47) from the Worbarrow succession (WOR 09), and a possible very early tricolpate (Fig. 9.47) from Isle of Wight Atherfield (BRN 126). CfA Barremian-*ring* from Isle of Wight Upper Compton (UCB 44) is again close (Fig. 9.48) to *Tucanopollis crisopolensis* (of Regali 1989), and finally an enigmatic tectate pollen from Isle of Wight Sandown (SYB 24).

Most of these records confirm the impression of clear radiations within the MCT group in Phase 4, although with one possible exception there is as yet no appropriate megafossil support. In June 1992, C.R. Hill of the British Museum (Natural History) announced at a meeting the discovery of numerous specimens in the Barremian

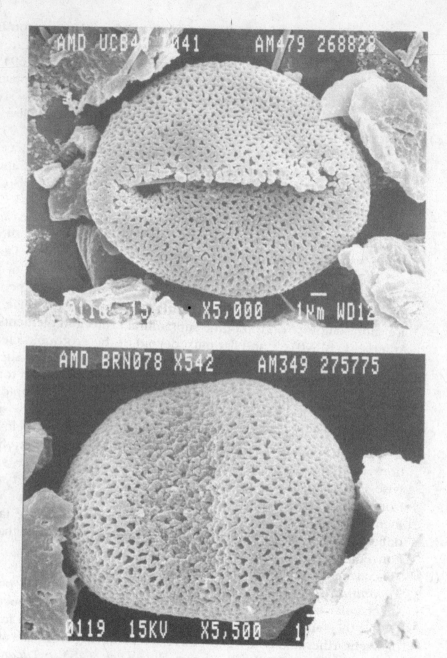

Fig. 9.34. Phase 4. SEMG of Retisulc group, Isle of Wight; CfA Retisulc-*monbac*, Upper Compton succession (0116); CfA Retisulc-*monbac*, Atherfield succession (0119). For notes, see explanation of Fig. 9.30.

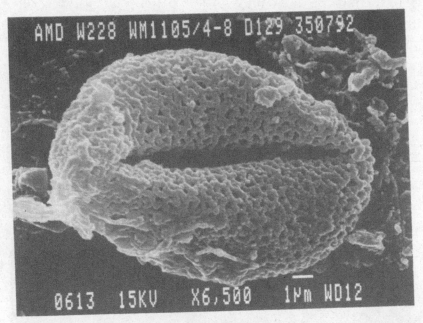

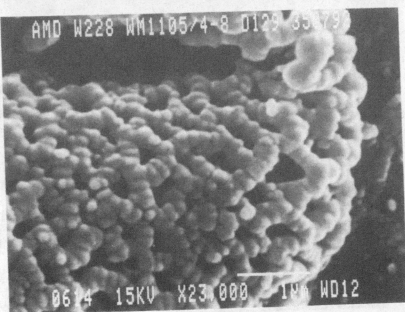

Fig. 9.35. Phase 4. SEMG of Retisulc group, Warlingham Borehole: CfB Retichot-*baccat* (0613) and enlargement (0614). For notes, see explanation of Fig. 9.30.

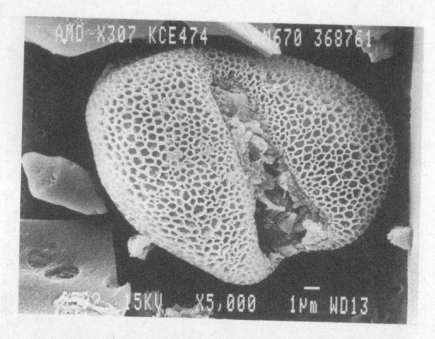

Fig. 9.36. Phase 4. SEMG of Retisulc group: Kingsclere Borehole, Paleotaxon Retisulc-*crochet* (0592); Atherfield succession, CfB Retisulc-*crochet* (0125). For notes, see explanation of Fig. 9.30.

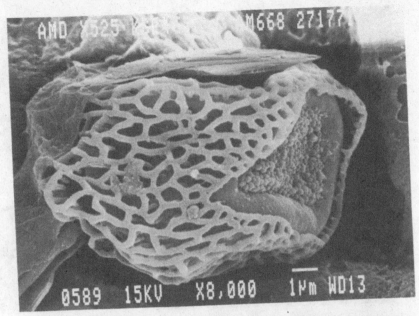

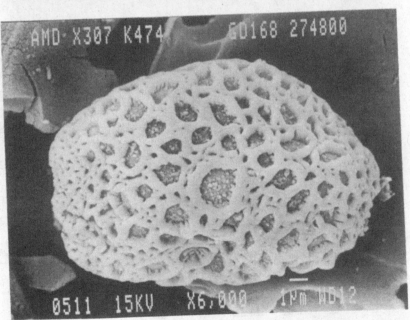

Fig. 9.37. Phase 4. SEMG of Retisulc group, Kingsclere Borehole: CfA Retisulc-*crochet* (0589); CfC Retisulc-*crochet* (0511). For notes, see explanation of Fig. 9.30.

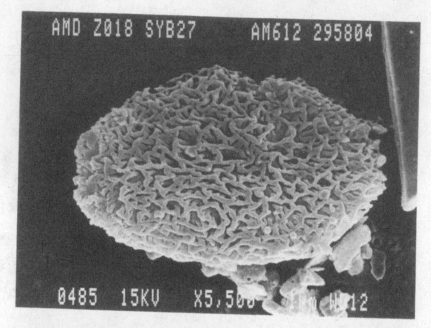

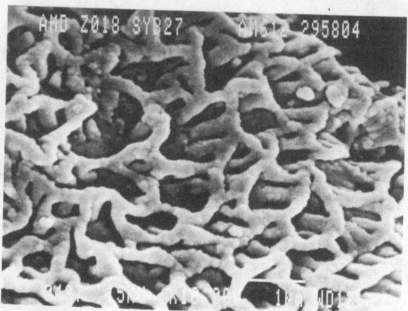

Fig. 9.38. Phase 4. SEMG of Retisulc group, Isle of Wight: CfA Aptian-*longcol* (0485); detail (0487). For notes, see explanation of Fig. 9.30.

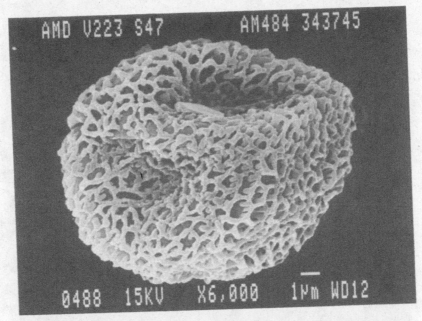

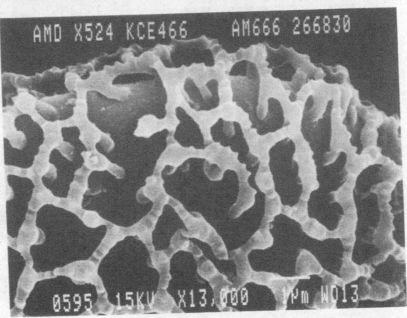

Fig. 9.39. Phase 4. SEMG of Retisulc group: CfA Aptian-*longcol* (0488), from Loc.S47; CfB Aptian-*longcol* (0595), Kingsclere Borehole. For notes, see explanation of Fig. 9.30.

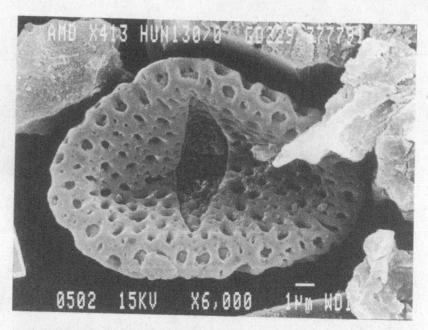

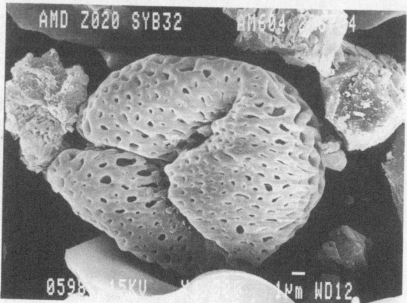

Fig. 9.40. Phase 4. SEMG of Perfotect group: Hunstanton Borehole, CfA Aptian-*perfotect* (0502), monosulcate; Isle of Wight, CfB Aptian-*perfotect* (0598), trichotomosulcate. Notes on photographs: observer identification, sample number, preparation number, stub number, stub-grid co-ordinates. Bottom line: photoserial number (Jeol JSM820), KV, original magnification, scale bar value, reference number. Other data are filed in the Department of Earth Sciences, University of Cambridge.

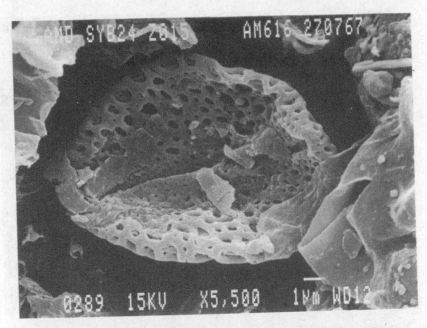

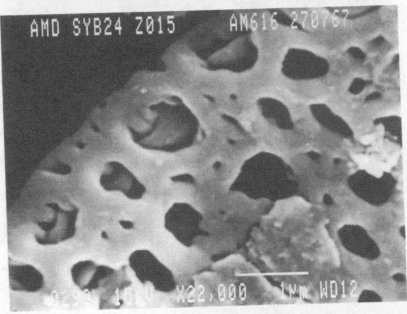

Fig. 9.41. Phase 4. SEMG of Perfotect group, Isle of Wight: CfA Aptian-*perfotect* (0289); detail with columellae (0290). For notes, see explanation of Fig. 9.40.

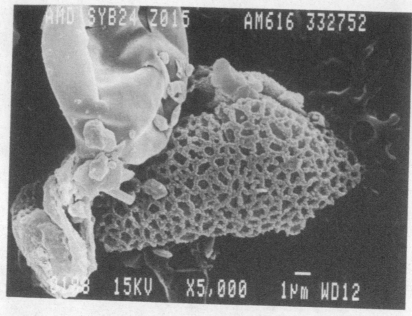

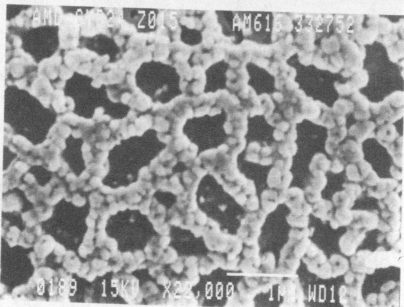

Fig. 9.42. Phase 4. SEMG of Retisulc group, Isle of Wight: Paleotaxon Aptian-*frill* (0188); detail of muri (0189). For notes, see explanation of Fig. 9.40.

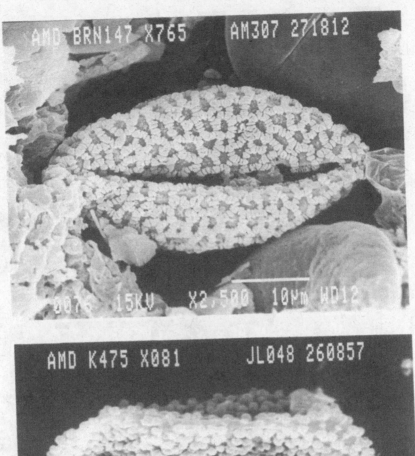

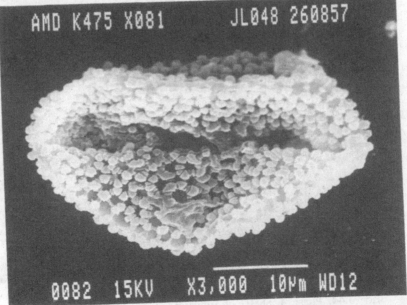

Fig. 9.43. Phase 4. SEMG of Superret group: Isle of Wight, CfA Superret-*croton* (0076); Kingsclere Borehole, Paleotaxon Superret-*croton* (0082). For notes, see explanation of Fig. 9.40.

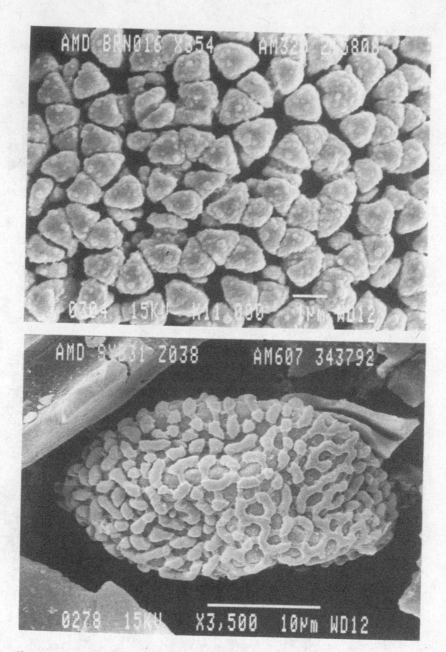

Fig. 9.44. Phase 4. SEMG of Superret group, Isle of Wight: CfB Superret-*triang* (0304), additional sculpture; CfC Superret-*croton* (0278), fused elements on muri. For notes, see explanation of Fig. 9.40.

Weald Clay of Surrey, England, of a new herbaceous seed plant. No flowers, *in situ* pollen or cuticle have yet been found but at least such a plant is urgently needed in explanation of the occurrence of some of the MCT group pollen. *Ephedripites* group pollen of the same small size as in earlier phases, continues at about the same level (Fig. 9.49), but a surprising specimen from Isle of Wight Upper Compton (UCB 40) shows an aperture comparable with that of many MCT pollen, filled with small irregular mural fragments; this and the variety of taxa within this group suggest that there may be no close affinity between Cretaceous '*Ephedripites*' and *Ephedra* pollen (of the living Gnetales).

Phase 5: consolidation. The final phase considered here includes the records of assemblages in which for the first time the overall quantity of angiosperm and angiosperm-like pollen reaches 10% of the whole palynomorph count in some samples, representing a significant step towards ultimate angiosperm dominance in Late Cretaceous time. The rocks concerned are of Early Aptian age and are best represented in the Isle of Wight Atherfield succession, east of Atherfield Point; the reference sample (F317) is from the 'Crackers Group III' of the Atherfield Clay, dated by the forbesi ammonite zone that lies immediately above the standard Aptian marine fissicostatus ammonite zone (mentioned above under Phase 4), which includes the Perna Bed and now also certain rocks below this. These sandy clays of Phase 5 contain both ammonites and dinoflagellates and only also contain so much pollen through paleogeographic accident; they were formed immediately south of a local Dorset–Isle of Wight high, which still provided adequate terrigenous material at a time when all other successions to the east and north were fully marine, as were also those stratigraphically higher in the Atherfield Clay. The Worbarrow succession of Dorset is comparable for similar reasons, although the detailed lithology is slightly less favourable (samples WOR41 to WOR38) for the recovery of small pollen.

Prominent new types of pollen include CfA Retisulc-*dubdent*, with narrow muri, large lumina (Fig. 9.50), short columellae and a tendency for detachment of the tectum from the nexine. CfA Retisulc-*dident* has large numbers of crowded very small lumina (Fig. 9.51). Paleotaxon Afropol-*murigroove* was described by Penny (1989) from the Hunstanton Borehole (HUN 130) as the first occurrence of such pollen (Fig. 9.52) in western Europe. Crotonoids are fewer than in Phase 4 and slightly different, as indicated (Fig. 9.53) by CfA Superret-*croton* from Isle of Wight Atherfield (F317) and a CfB record of the same from Hunstanton Borehole (HUN 120); in addition (Fig. 9.54) CfA Superret-*krinkel* from Isle of Wight Atherfield (F317). A further single tricolpate grain (Fig. 9.55) from Isle of Wight Atherfield (sample BRN 094) can easily be ignored but may also be comparable with '*Tricolpopollenites*' *crassimurus* Groot and Penny, as employed by Doyle. Various well-preserved enigmatic

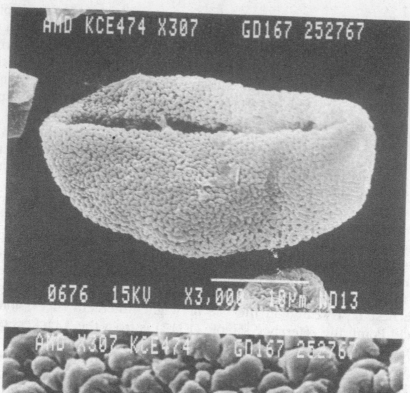

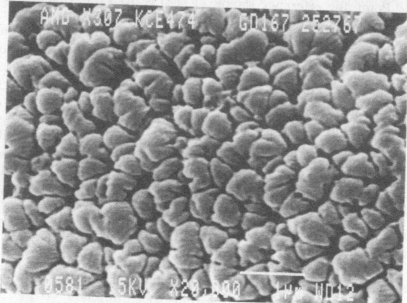

Fig. 9.45. Phase 4. SEMG of Superret group, Kingsclere Borehole: Paleotaxon Superret-*subcrot* (0676), detail (0581). For notes, see explanation of Fig. 9.40.

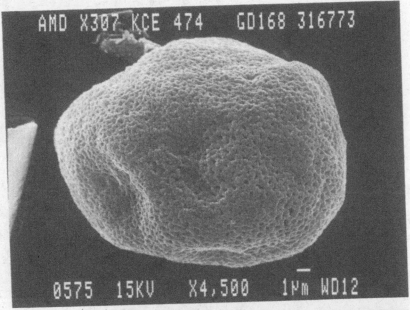

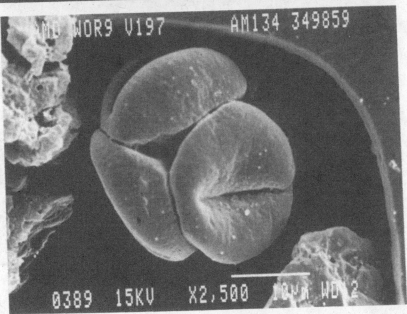

Fig. 9.46. Phase 4. SEMG of Superret group, Kingsclere Borehole: CfA Superret-*subcrot* (0575); CfA *Eucommiidites minor* tetrad (0389). For notes, see explanation of Fig. 9.40.

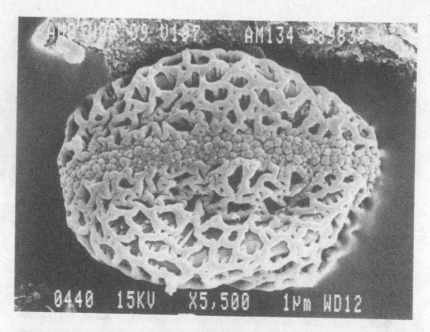

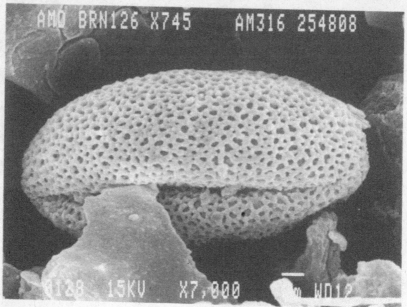

Fig. 9.47. Phase 4. SEMG of Retisulc group, Worbarrow succession, CfB Aptian-*longcol* (0440); possible tricolpate pollen (0128), Isle of Wight. For notes, see explanation of Fig. 9.40.

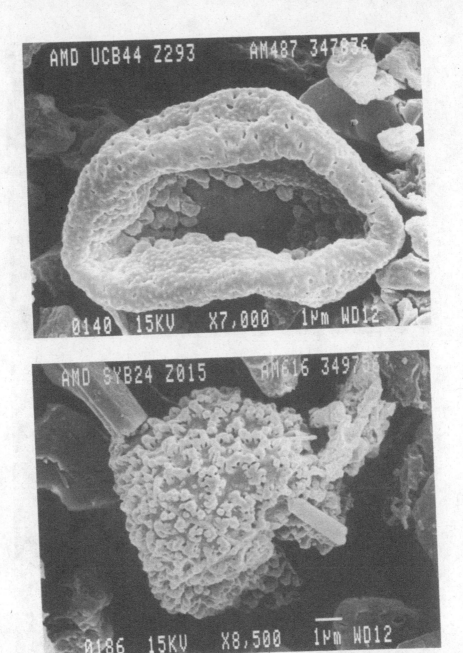

Fig. 9.48. Phase 4. SEMG, Isle of Wight: CfA Barremian-*ring* (0140); CfC Superret-*croton* (0186). For notes, see explanation of Fig. 9.40.

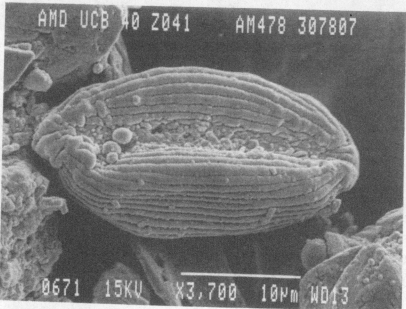

Fig. 9.49. Phase 4. SEMG of Ephedripites group, Isle of Wight: CfA Barremian-*murifrag* (0669), Paleotaxon Barremian-*murifrag* (0671). For notes, see explanation of Fig. 9.40.

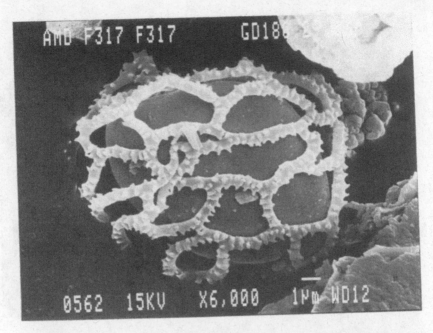

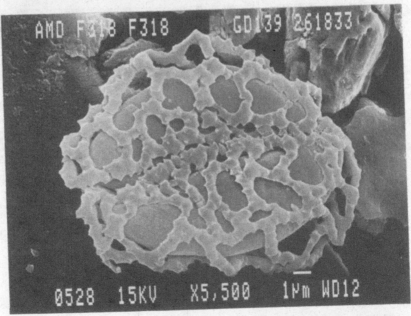

Fig. 9.50. Phase 5. SEMG of Retisulc group, Isle of Wight: CfA Retisulc-*dubdent* (0562); CfB Retisulc-*dubdent* (0528). Notes on photographs: observer identification, sample number, preparation number, stub number, stub-grid co-ordinates. Bottom line: photoserial number (Jeol JSM820), KV, original magnification, scale bar value, reference number. Other data are filed in the Department of Earth Sciences, University of Cambridge.

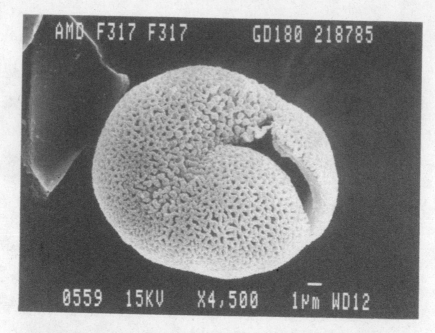

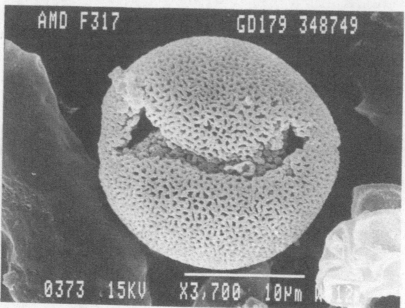

Fig. 9.51. Phase 5. SEMG of Retisulc group, Isle of Wight: Paleotaxon Retisulc-*dident* (0559 and 0373). For notes, see explanation of Fig. 9.50.

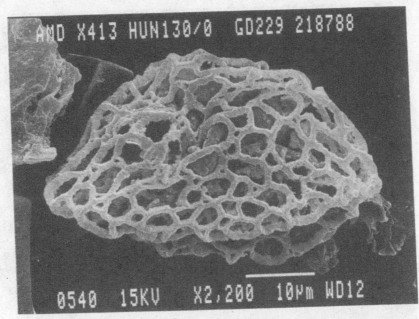

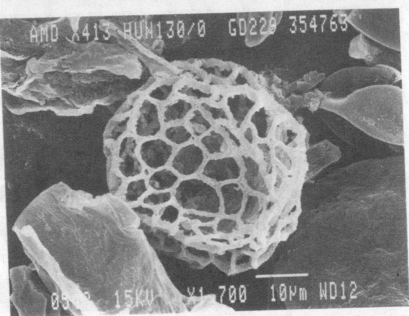

Fig. 9.52. Phase 5. SEMG of Afropollis group, Hunstanton Borehole; Paleotaxon Afropol-*murigroove* Penny (0540 and 0542). For notes, see explanation of Fig. 9.50.

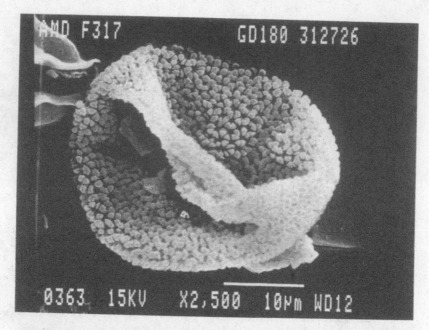

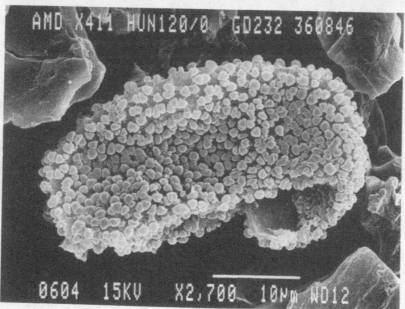

Fig. 9.53. Phase 5. SEMG of Superret group: Isle of Wight, CfA Superret-*croton* (0363); Hunstanton Borehole, CfB Superret-*croton* (0604). For notes, see explanation of Fig. 9.50.

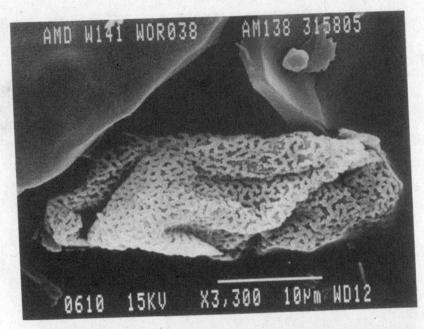

Fig. 9.54. Phase 5. SEMG of Superret group, Worbarrow succession: CfA
Superret-*krinkel* (0610); detail (0611). For notes, see explanation of Fig. 9.50.

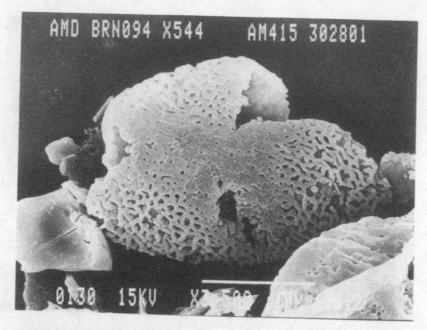

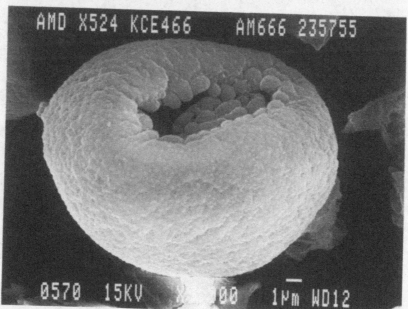

Fig. 9.55.   Phase 5. SEMG: Isle of Wight, CfB *'Tricolpopollenites' crassimurus* (0130); Kingsclere Borehole, CfC Hauterivian-*cactisulc* (0570). For notes, see explanation of Fig. 9.50.

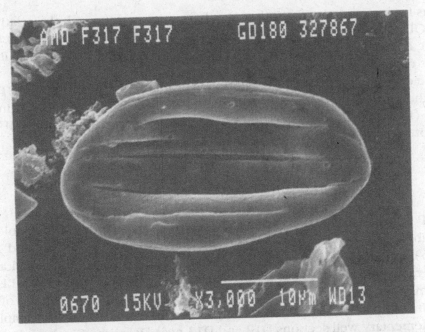

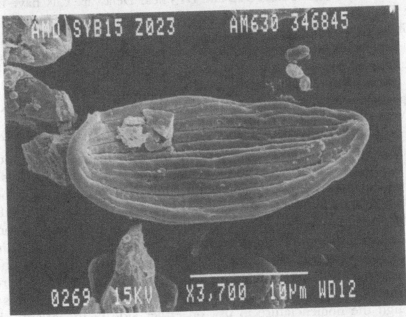

Fig. 9.56. **Phase 5. SEMG of Ephedripites group, Isle of Wight; CfA Barremian-**
fourboat (0670); CfA Aptian-*monotwofour* (0269). For notes, see explanation of
Fig. 9.50.

grains form a reminder that other taxa have persisted through the relative flood of MCT pollen into Phases 4 and 5, e.g. CfC record of a small monosulcate (Fig. 9.55). *Ephedripites* pollen continues (Fig. 9.56) but does not increase in quantity or diversity at this level.

The criterion for Phase 5 recognition is primarily the quantity of MCT pollen, and therefore the presence of Paleotaxon Retisulc-*dubdent* records in Kingsclere (KCE 469 to KCE 466) and of CfA *Afropollis* in Hunstanton (HUN 130) do not serve to take those samples out of Phase 4 into Phase 5.

9.14   Comparison of other successions: North America

The Potomac Group succession of eastern North America, palynologically described first by Brenner (1963) and subsequently by Doyle and Hickey (1976) and Doyle (1982, 1983), is essentially of Aptian–Cenomanian age; the stratigraphy was well set out by Doyle and Hickey (1976, p. 145) and is based on a number of separated outcrops yielding megafossil leaves, many of which are difficult to correlate. Abundant palynomorphs from two closely sampled and complementary well sections D12 and D13 near Delaware City have been used to date the outcrop occurrences in Virginia and Maryland. Zone I of the Potomac Group, covering the Patuxent Formation below and the Arundel Clay above, represents the earliest fossils obtainable from the lowest Cretaceous sediments available above a major regional unconformity over Triassic and older rocks. The zone has been said by some to be of Barremian–Aptian age but from a comparison with the English material of the MCT-type pollen described by Doyle and Hickey (1976), including those from the lowest D12-770 feet samples, the oldest American pollen appears to belong to the English Phases 4 and 5 (Fig. 9.57) and must be all Aptian if the paleomagnetic alteration to the Isle of Wight Atherfield succession is accepted (see Section 9.3). This comparison could obviously be sharpened with equivalent scanning electron micrographs (SEMG) on both sides; Walker and Walker (1984, figs. 7–11) illustrated very clearly their version of *Clavatipollenites hughesii* Couper from Lower Zone I and Aff. *Retimonocolpites* Pierce sp. 2 from Upper Zone I, which are in agreement with the results above (Fig. 9.57). Unfortunately all the other grains they illustrate equally well, including *Retimonocolpites dividuus* Pierce, were taken from Upper or Middle Zone IIB and are thus not relevant in this context.

Although the nomenclature in use on either side of the Atlantic bears no resemblance, and although light microscope and SEM studies are so different, both sides are satisfied with the comparisons made above. While in no way claiming that the English taxonomy employed here (Hughes 1989) is alone satisfactory, there are obvious difficulties in trying to continue to make easily retrievable records within the three genera originally employed by Doyle and others: *Clavatipollenites* Couper 1958, as explained elsewhere, is now unsatisfactory because the holotype of the type species cannot be closely identified or

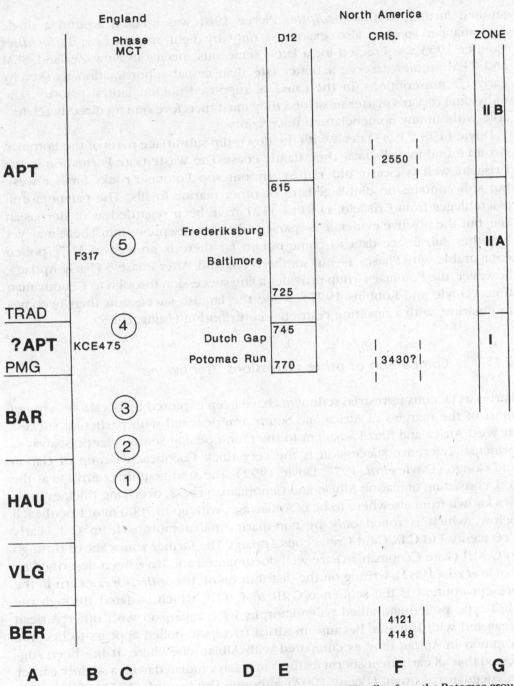

Fig. 9.57. Comparison of occurrence records of MCT pollen from the Potomac group of eastern North America (Doyle and Hickey 1976) with those from the English successions. A, ages of the Geologic time-scale (GTS), with alternative positions of the Barremian/Aptian boundary. TRAD, at the Perna Bed; PMG, modification from paleomagnetic evidence; B, English sample numbers for reference; C, MCT Phases 1–5 in England; D, Potomac outcrop localities; E, depths in feet of D12 Borehole (Delaware City); F, depths in feet of Crisfield Borehole; full line, cores; broken line, cuttings; G, zones in use for Potomac succession.

studied further; *Retimonocolpites* Pierce 1961 was erected around a single Cenomanian species, also examined only by light microscopy; *Liliacidites* Couper 1953 was erected for a late Cretaceous species in New Zealand. SEM and TEM studies deserve a better fate than casual subordination to sketchy early circumscriptions, in the cause of supposed nomenclatural priority. The individual records matter most, and they must therefore remain directly retrievable without any nomenclatural uncertainty.

Doyle (1982, 1983) dealt with the down-dip subsurface parts of the Potomac Group found in wells near the Atlantic coast. The Waste Gate Formation in the Crisfield well is clearly older than any outcrop Potomac rocks further west, but still contains no dinoflagellates or other marine fossils. The palynomorph assemblages from Crisfield 4148 to 4121 have been regarded as of Berriasian age, but the positive evidence is sparse and rather disappointing. There may yet be other subsurface data to come but so far there is no sign of MCT pollen comparable with Phases 1–3 in southern England. After Phase 5 (Early Aptian), however, the Potomac Group provides a fine succession through to Cenomanian time (Doyle and Robbins 1977), while the English succession then becomes too marine, with a resulting restricted contribution (Laing 1975).

9.15   Comparison of other successions: Gondwana

Early Cretaceous terrestrial sediments have been explored by boreholes around most of the margins of Africa and South America, and with particular success in West Africa and Brazil adjacent to the then-opening South Atlantic seaway; a principal reference succession is the very thick Cocobeach Group of Gabon and Congo (Doyle *et al.* 1977, Doyle 1992). The stratigraphic control is at the top, consisting of marine Albian and Cenomanian rocks, overlying thick evaporites known from elsewhere to be of Aptian age, with up to 7000 m of Cocobeach below, which is zoned only by non-marine palynomorphs from C I (Early Cretaceous) to C IX (Mid-Cretaceous Aptian). The further zones above through to C XIII (Late Cenomanian) are well documented and have been described by Doyle *et al.* (1982), writing on the distribution of *Afropollis*. Of concern in the present context is the sequence C III to C VIII, which is dated through the MCT-type pollen and allied palynomorphs by comparison with other African areas and with Laurasia. Because in Africa tricolpate pollen appears to become common in Aptian time as compared with Albian elsewhere, it has been suggested that all early angiosperm features may have radiated from a slightly earlier paleoequatorial origin (Penny 1992), although Doyle *et al.* (1977, 1982) were properly cautious about this. As indicated in Fig. 9.58, Zone C V could be taken to represent the same time as the Phase 3 assemblage in England of Barremian age, with Zone VI having innovations similar to those of Phase 4 and therefore probably of Aptian age. Zone C IV, below, appears to have no independent distinguishing character, and *Dicheiropollis etruscus* of Zones III and II has

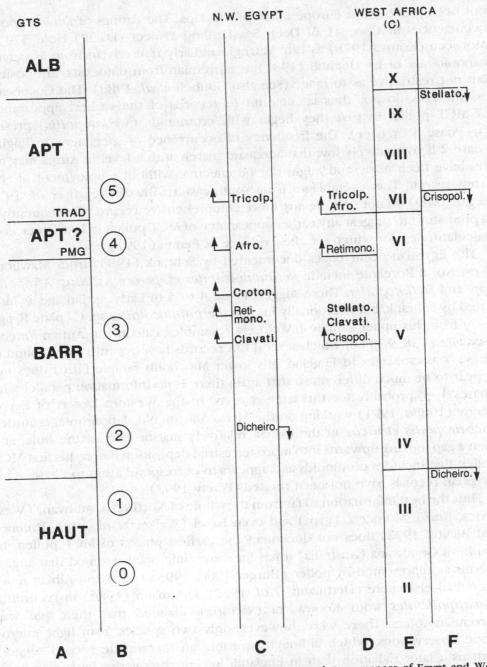

Fig. 9.58. Interpretation of the MCT palynomorph occurrences of Egypt and West Africa. A, ages of the geologic time-scale, with Barremian/Aptian boundary as in Fig. 9.57; B, MCT Phases 1–5 in England; C, entry of fossil groups in north-west Egypt (Penny 1988a, 1991, 1992); D, entry of fossil groups in Gabon, West Africa (Doyle et al. 1977, 1982); E, C zone division of Gabon succession; F, last occurrences of certain palynomorphs: Dicheiro., *Dicheiropollis etruscus*; Clavat., *Clavatipollenites* group; Crisopol., *Inaperturopollenites crisopolensis*; Stellato., *Stellatopollis* group; Retimono., *Retimonocolpites* group; Croton., crotonoid group; Afro., *Afropollis* group; Tricolp., *Tricolpites* group.

not been recorded in Europe north of the Alps. The datings of *Dicheiropollis* occurrences in Core 41 of Deep Sea Drilling Project (DSDP) Hole 370 off Morocco (Kotova 1978) as Late Valanginian/Early Hauterivian from associated nannofossils, or by Hochuli (1981) as Barremian from dinocysts, are positive but not restrictive as to range (see also Goubeli *et al.* 1988). The Cocobeach assemblages do not thus far amount to records of the earliest appearances of MCT pollen in that they begin with crotonoids (*Stellatopollis*) present (= Phase 3 or later). The frequency of occurrence of specimens in English Phase 2 is so relatively low that adequate search at that level in Africa may not yet have been made and might not be effective without employment of SEM strew-search. The assemblages from south-eastern Brazil (Regali *et al.* 1974, Regali 1989) do not provide any more comprehensive record. Consequently it is premature to suggest any earlier appearance of MCT pollen in low Cretaceous paleolatitudes, for which see the evidence of Penny (1992).

The Egyptian occurrences documented by Schrank (1983) from Mawhoub West No. 2 Borehole include *Retimonocolpites* cf. *peroreticulatus, Afropollis* spp. and *Stellatopollis*. These suggest Phase 4 to 5 of Early Aptian age as indicated by Schrank, but additionally his new *Retimonocolpites* sp. C (plate 8, figs. 4–7) from his upper sample IIA closely resembles Paleotaxon Aptian-*longcol* (see Figs. 9.38, 9.39) of which several CfA records have recently been found in Phase 4 assemblages in England. His lower Mawhoub sample (IIIc) does not appear to be much different, so that again there is no information parallel with Phases 1–3, probably for the same reasons. In the Western Desert of Egypt, Penny (1988*a*, 1991) working on the Mersa Matruh No. 1 Borehole, recorded *Dicheiropollis etruscus* in the lowest relatively marine part of the hole, and then a gap moving upwards into more terrestrial deposition before his first MCT pollen that includes crotonoids and appears to correspond again to Phase 4; his Egyptian records have not been revised (Penny 1992).

Thus the best information so far from the whole of Northern Gondwana (West Africa, Brazil, Morocco, Egypt) and even Israel (Walker *et al.* 1983, Brenner and Bickoff 1992) does not document the earliest phases of MCT pollen. In Southern Gondwana (Australia) it has for some time been assumed that angiosperms as represented by pollen (Burger 1980, 1993) entered in Albian time, later than elsewhere (Dettmann *et al.* 1992). Dettmann (1986) in recording *Clavatipollenites* from Koonwarra (Victoria), claimed that their age was Barremian/Aptian. There were, however, only two species from light microscope observations, which although probably an incomplete record suggest perhaps a later radiation than in England. Recent discoveries have also been made in India (Tripathi and Tiwari 1991).

Finally, a detached anther (600 μm x 250 μm), with *Clavatipollenites*-type pollen, has now been recorded from the Early Aptian Baguero Formation of Argentina (Archangelsky and Taylor 1993). The MCT pollen grains concerned closely resemble those of Aptian-*frill* of Phase 4 (see Fig. 9.42) and not, as cited by these authors, the grains figured by Hughes (1977), which occur in Phase 2 (Early Barremian) in England.

9.16   Scope of present MCT study

From the twelve relevant English successions mentioned above, about 600 samples were collected or selected and of these 550 were processed to slide preparations. Approximately 350 sets of stubs were made from these preparations and submitted to search by SEM, almost always for at least an hour each and frequently for very much longer time with the more successful stubs. Virtually all this observation work was done successively by John F. Laing during one year, Gillian E. Drewry for six years, and Audrey B. MacDougall part-time for ten years; additionally one set of important records from the Warlingham Borehole was added by Ian C. Harding, who also checked the dinocysts encountered by the other observers. As recorded above, most of the SEM stubs made seventeen years ago are still usable without faults. In all, some 2200 specimens of MCT pollen (certain and probable) and '*Ephedripites*' were recorded and photographed, excluding dinocysts, *Classopollis* and other pollen, and spores. The work continues at present on sample preparations of Berriasian–Hauterivian age in search of forerunners of the MCT pollen. Stub-grid co-ordinates have been recorded for all specimens, so that re-examination is easily arranged as necessary. All documents, photonegatives, stubs, slide preparations and samples are deposited with the Sedgwick Museum, Cambridge. Most specimens are recorded in new style taxa (Paleontologic Data-handling Code of Hughes (1989)) and comparison records thereto. The relevant traditional taxa were all erected three decades ago for light microscope work only and have in most cases been downgraded or confused by subsequent use and many emendations, but the new-style taxa are binominally compatible with the old and the two can thus be used together whenever necessary. The singularity of MCT pollen is its small size, which results in its being missed in most reconnaissance work and necessitates an SEM/TEM observation base for making usable records. The older combination of light microscopy and ICBN-driven taxonomy has reached a level in this field beyond which little improvement or further enlightenment can be expected.

9.17   Regional palynologic characters

Early Cretaceous paleolatitudinal differences of palynologic assemblages were to be expected and have been demonstrated in distinctions between Laurasia and Northern Gondwana at about 25–30° Hauterivian N. Large fern spores and bisaccate pollen were characteristic of Laurasia and had almost disappeared in Northern Gondwana by Barremian time: *Dicheiropollis etruscus* was virtually restricted to Northern Gondwana (including what is now Italy). The distribution of the palynomorphs new in Barremian/Aptian time, such as MCT pollen including both *Afropollis* and *Inaperturopollenites crisopolensis*, appears not to have been controlled by latitude; this suggests that the paleolatitude climatic gradient may have become slight, and that time-correlation potential of these fossils may thereby have increased.

9.18   Botanical affinity of MCT pollen

The best possibility so far is the new plant mentioned above and not yet formally described, from the Weald Clay of Surrey. The principal pollen type is relatively simple (quoted usually as *Clavatipollenites*) but is by no means unique to the frequently quoted living family Chloranthaceae, which is mostly from the margin of the current western Pacific Ocean but also includes *Hedyosmum* from the Caribbean area. This wide pantropical and subtropical distribution does suggest a possibly widespread set of relics of a former more universally occurring family, but the family is heterogeneous and without other megafossil record. There is a possibility that ultimately the Chloranthaceae may be shown to have been descended from MCT ancestors, but no connection is worth showing until some proof of intermediate fossils has been provided, and the name should not be used in any hypothesis without the weakness of the evidence being emphasised. Other MCT 4 pollen includes the forms claimed by Doyle (1973) as being of probable monocotyledon origin (see Chapter 11). Crotonoid exine sculpture appears to be very distinctive for such early occurrence as MCT 3 and 4; however, the patterns are known from living Buxaceae, Thymeliaceae and Euphorbiaceae (Erdtman 1952), and even Proteaceae, all with different aperture arrangements. Trincao and Pais (1992) referred to a fossil example under the name of *Portucallipollis* from Berriasian rocks of Portugal, but the grain-size diameter of 65 μm suggests some doubt concerning such an affinity. Unfortunately none of these points positively assists with an explanation of any of the numerous patterns of MCT pollen awaiting further attention.

10 First convincing angiospermous fossils

10.1 Stratigraphic considerations and limitations

In Cretaceous Cenomanian time, records of the first generally convincing angiosperm flower megafossils, supported in general by subsequent radiations, have been made in widely separated localities. From Late Albian times, just earlier, there are numerous more fragmentary megafossils that provide some hope of eventual success in that age too. Triaperturate palynomorphs become regular members of assemblages in the Albian age (Herngreen and Chlonova 1981, Penny 1988b, 1992). Unfortunately stratigraphic circumstances are not particularly favourable either in taphonomy or in correlation, so that no major, single, well-preserved megafossil flora is available from these ages. Although many smaller floras of western North America are listed by McClammer and Crabtree (1989), some of the best-preserved fossils come from Kansas and Nebraska in the Dakota Formation of Early Cenomanian age, but that succession is not favourably continuous for plant fossil occurrences either upward or downward. In the Potomac Formation of Virginia and Maryland there is a much longer and generally favourable succession, with many leaf fossils through from Aptian age (Zone I of Brenner 1963) rocks below, to Cenomanian (Zone III) and beyond; because the regional dip is very low and as the sediments are unconsolidated and make poor outcrops, local stratigraphic correlation is difficult to achieve.

The marker points and type successions for this part of the stratigraphic scale have been selected in western Europe so that difficulty in correlating with marine fossils increases away from that region (Fig. 10.1). In Europe from Aptian time onwards the sedimentation was becoming rapidly more marine and calcareous upwards towards the widespread nannoplankton abundances and dominances of the Chalk Sea. An important modification was in Bohemia (western part of Czechoslovakia), where Mid-Cenomanian clastic sediments marginal to the European Chalk Sea provide a well-known flora (Velenovsky and Viniklar 1926–31, Pacltova 1971, 1977), although there is no appropriate Albian succession below in that area (Fig. 10.2).

In Asia small mid-Albian leaves and other fossils have been recorded from Kazakhstan and are probably the earliest of their kind. Fossils of mid-Albian age from Kolyma in the Siberian Far-East lack cuticle through pyroclastic and possibly other influences but include various forms not recorded elsewhere.

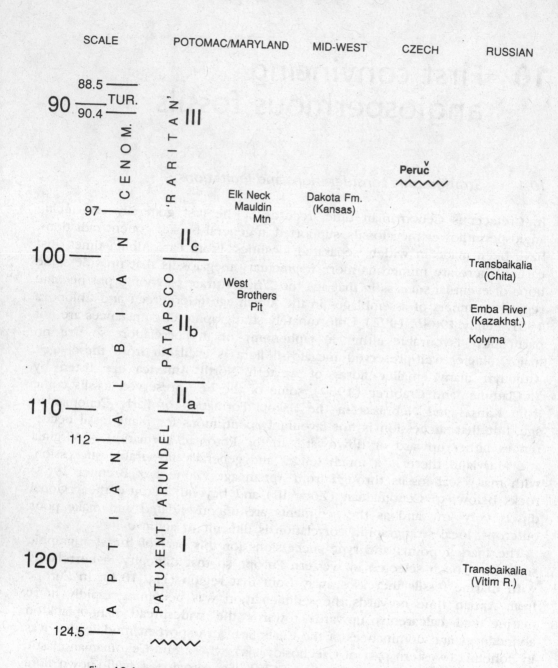

Fig. 10.1. The stratigraphic position of the principal localities mentioned in this chapter, showing the relative durations of the ages Cretaceous Aptian to Turonian inclusive.

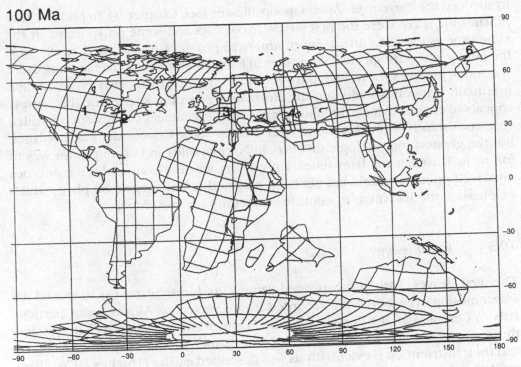

Fig. 10.2. Paleocontinental map for Cretaceous Late Albian time, 100 Ma, to indicate the paleolatitudes of the principal localities of mid-Cretaceous mega- and mesofossils of angiospermous plants. 1, Kansas; 2, Maryland; 3, Bohemia; 4, Emba River; 5, Transbaikalia; 6, Kolyma (map courtesy of A. G. Smith, 1992).

Although the stratigraphic coverage indicated above for this critical time in land plant history is far from complete, it is not unusually weak for a single age selected at random from any part of the Mesozoic time-scale of the world. Additionally the recent very successful recording of mesofossils (e.g. Crane *et al.* 1986, Friis *et al.* 1988, Drinnan *et al.* 1991) has now provided more potent information in this angiosperm context than would have come from a standard megafossil flora.

10.2   *Palynologic background*

From Albian and Cenomanian rocks there has been a pattern of many records in most paleolatitudes from 75° Cretaceous N to 60° S (Herngreen and Chlonova 1981); subsequent records to date emphasise and consolidate the pattern (e.g. Juhasz and Goczan 1985, Herngreen and Duenas Jimenez 1990, Hua 1991, Shang and Wang 1992, Spicer *et al.* 1992). There appears to be general agreement that there was poleward migration from low paleolatitudes in which *Retitricolpites* in Early Albian time was the typical first triaperturate pollen (Penny 1992) to succeed the Barremian–Aptian monosulcates (see Chapter 9). In Mid and Late Albian time there were the first striate tricolpates and some polyporates, at the same time as the ASA (Africa–South America) province was flooded briefly with the unique elaterate species. Thereafter in Early and Mid-Cenomanian time there was much greater variety, including very small tricolpates, more polyporates and eventually triporates including the Normapolles group; at present this variety appears to exceed that credited to megafossils. Percentages of angiosperm pollen rose consistently through time to over 50% in some Cenomanian assemblages, but the greatest morphologic diversity in Normapolles and other groups was to follow in Turonian and later times. Only a very small selection of the numerous dispersed pollen available has yet been matched to any megafossil plant, which emphasises the incentive to continue searching for mesofossils.

10.3   Fossil leaves

The first leaves with an expanded lamina and reticulate (as opposed to anastomosing) venation were recorded from Cretaceous Albian strata, particularly in Central Asia, the Russian Far-East, Portugal and the eastern United States; the earliest ones were very small (<2 cm diameter) and very simply veined and their distribution is essentially as was described earlier (Hughes 1976, chapter 10). Hickey and Doyle (1977) presented the same facts more clearly and their diagrams have been quoted extensively. Unfortunately the Russian occurrences yielded no cuticle, the stratigraphic control for the early Portuguese collections was insecure, and the most significant early American specimens, although sometimes cuticle-bearing, came from localities no longer available. Because these leaves are small, more finds can be expected as stratigraphic subdivision of non-marine rocks improves. Most of the leaves have been given names that suggest attribution to extant taxa, but almost all these attributions are better ignored; interpretation of the vegetation (Upchurch and Wolfe 1987) can be very successfully made from leaf-form (Fig. 10.3) analysis and distribution regardless of the names given.

10.4   Albian records of reproductive organs

There are numerous scattered and mostly fragmentary records from all the regions in which leaves of Albian age were collected. Great variety is recorded from the mid-Albian of the Russian Far-East Kolyma Basin (Samylina 1968), but the preservation is adverse as with the leaves. Although both apocarpous and possibly syncarpous forms are recorded, they are incomplete as fossils and lack both perianth and cuticle, and there is no recoverable pollen.

Friis and Crepet (1987) provided useful comment on many of these fossils, although they also confuse by frequent use in description of adjectives such as

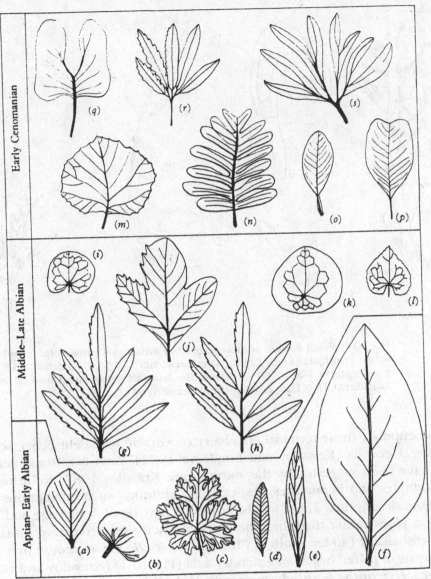

Fig. 10.3. Aptian–Cenomanian age types of leaf fossil, approximately ×1/2. (a) *Celastrophyllum obovatum*; (b) *Proteaephyllyum reniforme*; (e) *Rogersia*; (f) *Ficophyllum*; (g,h) *Sapindopsis*; (i) *Populophyllum reniforme*; (j) platanoid type; (q) *Liriophyllum*; (r) *Dewalquea* (from Upchurch and Wolfe 1987).

magnolioid and platanoid. These avoidable words conjure up additional characters that are *not* displayed by the fossils, and they also detract from the evidence of such observed characters as can be shown to be inappropriate for organisms bearing these names.

From the the Early Aptian–Albian rocks of the Baikal area (Vitim River) are recorded *Eoantha zherikhinii* consisting of a flower of four gynoecial valves each with an orthotropous ovule (Krassilov 1986), and *Baisia hirsuta* with a

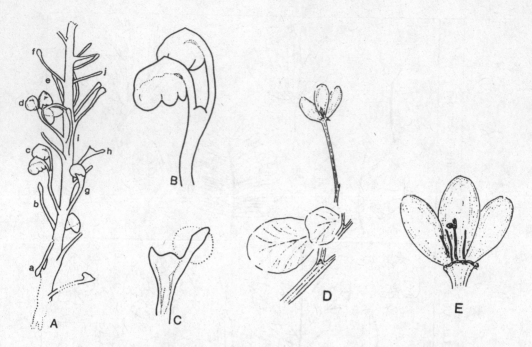

Fig. 10.4. Fossil flowers of Albian age from Russia, A–C, *Meeusella* from Late Albian strata, Transbaikalia near Chita; (A) inflorescence, ×2.5; (B, C) detail, ×8.5 (Krassilov and Bugdaeva 1988c); (D–E) *Hyrcantha karatscheensis*, Mid-Albian of Emba River, Kazakhstan; D, ×1.25; E, ×3.5 (from Krassilov *et al.* 1983).

single cupulate ovule terminal on a hairy receptacle with dehiscence scars suggesting a corolla (Krassilov and Bugdaeva 1982). Both fossils are very small and have been suggested by the authors (see Krassilov 1987) to be related to gnetophytes and bennettites, respectively, although such attributions do not help the description. These fossils, however, may be of Aptian age and thus 10 million years earlier than the others under discussion. From Late Albian rocks of Transbaikalia (40 km south of Chita) is *Meeusella proteiclade*, a new staminate organ of paired heads on a slender axis (Fig. 10.4) (Krassilov and Bugdaeva 1988c). *Hyrcantha karatscheensis* from Mid-Albian rocks of the Emba River of western Kazakhstan is an inflorescence of small flowers with robust carpels and delicate stamens (Fig. 10.4) on a trimerous or possibly pentamerous plan (Krassilov *et al.* 1983), and *Caspiocarpus paniculiger* from the same area is an inflorescence of follicles only with no stamens or perianth (Krassilov 1984, 1991); these two more closely foreshadow Cenomanian fossils such as *Prisca* described below (Retallack and Dilcher 1981b).

As in the case of the leaves, the North American fossils from Maryland are better preserved and have provided more detail through SEM study. Friis *et al.* (1988) described from the Late Albian West Brothers Pit of northern Maryland, in a preservation recorded as lignitic, a very small staminate flower with reticulate tricolpate pollen of less that 10 μm diameter but regarded it as 'chloranthoid'. From the same locality *Platananthus potomacensis* consists of staminate flowers in a dense inflorescence; the flowers have five stamens (Friis *et al.* 1988) and some protective tepals and yielded *Tricolpites minutus* (Brenner) Dettmann of diameter 9–13 μm with tectal lumina decreasing in size towards the apertures. *Platanocarpus marylandicus* is a pistillate inflorescence of a hundred small flowers on a 3–4 mm axis; the two inflorescences are similarly preserved and associated, as is also a rare palmate-lobed leaf; as the names imply, a Platanaceae affiliation has been assumed. *Spanomera marylandensis* from the same locality depends heavily for its interpretation on the Cenomanian *S. mauldinensis* (see below), which is much more completely known (Drinnan *et al.* 1991). The Albian fossils, although distinctive, are fragments with no complete inflorescences or even flowers, but the associated pollen *Striatopollis* closely resembles the dispersed *S. vermimurus*. Drinnan *et al.* mentioned other comparable mesofossils as yet undescribed from the West Brothers Pit, and from earlier rocks elsewhere. The Albian fossils are thus varied and very incomplete, and further fortunate discoveries of well-preserved mesofossils are urgently needed, but also with concomitant good stratigraphic control.

10.5   Cenomanian flowers

In three separate Laurasian regions, Kansas (Western Interior of North America), Maryland, and Bohemia (Czechoslovakia), much more complete and character-bearing fossils have been found than in the Albian rocks mentioned above. The flowers concerned can be divided into at least three distinct groups that will be labelled here, for continuity of discussion with recently published papers, in accordance with their supposed extant plant affinities as M (magnoliids), H (hamamelids) and R (rosiids) fossils.

The M fossil group includes *Archaeanthus linnenbergeri* (Fig. 10.5) from the Early Cenomanian Dakota Formation of Kansas (Dilcher and Crane 1984, Retallack and Dilcher 1986). This fossil consists of a long stout axis bearing many spirally arranged follicles with seeds, and also scars lower down, suggesting the abscission of stamens and perianth, although neither of these has been seen. *Lesqueria elocata* from the same succession but in less favourable preservation (Crane and Dilcher 1984) differs in possessing a much shorter compact receptacle but correspondingly less clear indication of abscission of other parts.

Perhaps relatable is *Triplicarpus purkynei* of Mid-Cenomanian age from the Peruc Formation of Bohemia (Czechoslovakia). The three follicles attached to a receptacle (Fig. 10.5) are unusually large (900 mm × 150 mm), although there is only one specimen (Velenovsky and Viniklar 1926–31). *Mauldinia mirabilis* from the Elk Neck Formation of Potomac Group III (Early Cenomanian) from north-eastern Maryland (Drinnan *et al.* 1990) is a lignitised lateral unit of an inflorescence with five slightly compressed flowers; the largest unit is 3.5 mm long but material is apparently abundant (although mostly fragmentary) and

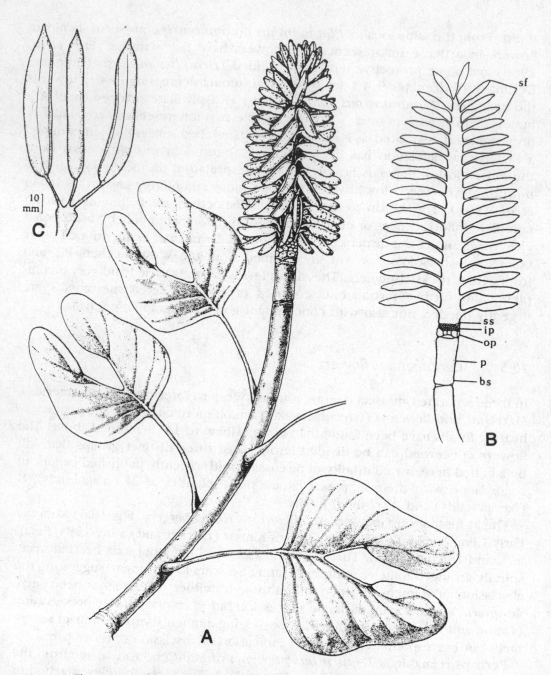

Fig. 10.5. Late Albian–Early Cenomanian *Archaeanthus linnenbergeri* from the Dakota Formation of Kansas. (A) Restoration of follicle bearing axis, with leaves of *Liriophyllum kansense*, ×1/2; (B) diagram illustrating the organisation; sf, stalked follicle; ss, presumed staminal scars; ip and op, perianth scars; bs, bud scale scars, (from Dilcher and Crane 1984); (C) *Triplicarpus purkynei*, ×1/2, from the Mid-Cenomanian Peruc Formation of Bohemia (from Friis *et al.* 1987).

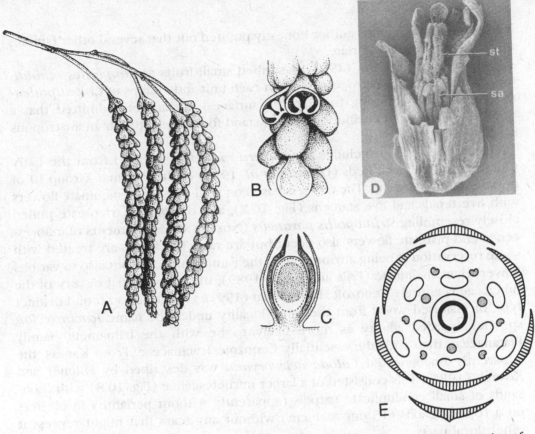

Fig. 10.6. Drawings of Cenomanian fossils. (A) *Prisca reynoldsii*, reconstruction of compression fossil from the Dakota Formation of Kansas, ×0.75; (B) two or more orthotropous ovules in each follicle, ×4; (C) single ovule with two integuments, longitudinal section, ×30 (from Retallack and Dilcher 1981*b*) (D) *Mauldinia mirabilis*, from the Elk Neck Formation of Maryland, broken flower showing tepals, stamens and staminodes (sa), ×14; (E) floral diagram of this fossil (staminodes stippled) (from Drinnan *et al*. 1990).

well-suited to SEM study and to the production of an interesting floral diagram (Fig. 10.6). *Mauldinia* is persistently attributed by its authors to the Family Lauraceae although comparison with extant genera is less than convincing and it is as well that the name selected for the fossil is local and neutral. *Prisca reynoldsii* (associated with a leaf *Magnoliaephyllum*) from the Dakota Formation of Kansas (Fig. 10.6) in a more usual but less favourable compression–impression preservation (Retallack and Dilcher 1981*b*) is believed to be very similar to *Mauldinia* (see also Kovach and Dilcher (1988) concerning a detached seed, probably of *P. reynoldsii*).

From the same locality as *Mauldinia* in the Elk Neck Formation of north-eastern Maryland, Herendeen (1991*a*) has described some small fusinised wood fragments under the name *Paraphyllanthoxylon marylandense* as also lauraceous, although this author honestly pointed out that several other families might be equally appropriate.

Finally Pedersen *et al.* (1991) described small fruits as *Couperites mauldinensis* with a single anatropous seed on each unit and clumps of *Clavatipollenites*-type pollen adhering to stigmatic surfaces; the authors admitted that a desired chloranthoid attribution cannot stand for this plant with an anatropous seed.

The H fossil group includes *Spanomera mauldinensis* also from the Early Cenomanian Elk Neck Beds (Drinnan *et al.* 1991) of the Potomac Group III of north-eastern Maryland. The taxon is based on well-preserved staminate flowers with five tepals and five stamens (Fig. 10.7), and the *in situ* tricolpate pollen closely resembling *Striatopollis paraneus* (Norris) Singh; fragments of inflorescence and pistillate flowers also occur but are rare. The fossils are treated with some reservation as being attributable to the Family Buxaceae but also to various 'lower' Hamamelididae (Friis and Crane 1989); the known fossil history of the Buxaceae is entirely Cenozoic. Herendeen (1991*b*) described a second distinct type of fusinised wood from the same locality under the name *Icacinoxylon* sp.; he regarded affinity as more likely to be with the hamamelid family Platanaceae than with the essentially Cenozoic Icacinaceae. From Kansas, the apparently unconnected *Caloda delevoryana* was described by Dilcher and Kovach (1986). This consisted of a larger infructescence (Fig. 10.8) with thousands of small conduplicate carpels (apparently without perianth) in clusters on a rope-like axis (15 cm × 1 cm) without any scars that might represent other floral parts.

The R fossil group is represented quite differently by compression preservation fossils of relatively small actinomorphic flowers from the Peruc Formation of Bohemia (Czechoslovakia). Krassilov and Pacltova (1989) redescribed *Asterocelastrus cretacea*, which is a pentamerous fruit with perianth members 10 mm long, with two anatropous seeds per carpel, ovary inferior, syncarpous (Fig. 10.9); cuticle was not obtained from the specimens. *Kalinaia decapetala* is a larger five-loculed syncarpous fruit (Velenovsky and Viniklar 1926–31) 15 mm in diameter with a ten-lobed perianth (Fig. 10.9); (see also Friis and Crepet (1987), who mentioned others, including less completely preserved examples from the Dakota Formation of Nebraska). These have all been very loosely referred to as rosiids (Krassilov and Pacltova 1989), but they are more importantly distinct and appear to have been the first of their kind.

10.6   Magnoliids and hamamelids

Terms such as these are deeply embedded in most descriptive work about mid- or late Cretaceous angiosperms. Class Magnoliopsida (Dicotyledones), Subclass Magnoliidae, Super-order Magnolianae, Order Magnoliales, and Family Magnoliaceae are all names of higher taxa that might be taken to be implied by the

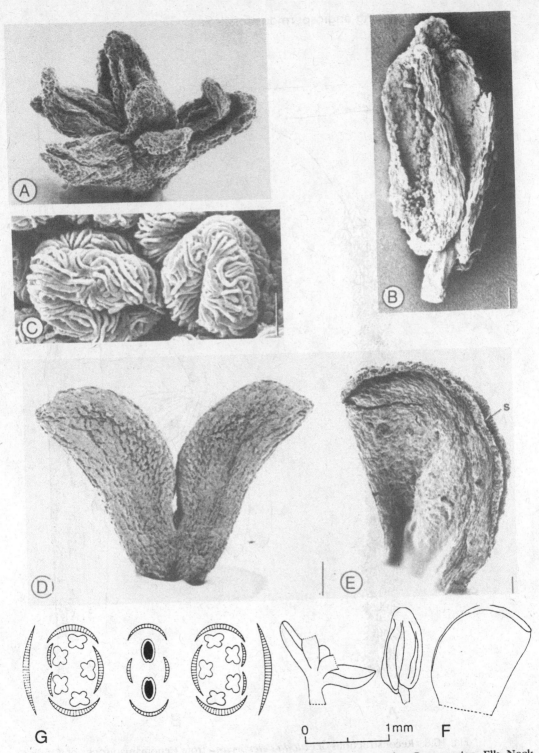

Fig. 10.7. Fossils of *Spanomera mauldinensis* from the Cenomanian Elk Neck Formation of Maryland. (A) Oblique lateral view of flower with tepals and stamens, scale bar 100 μm; (B) view of partly dehisced stamen, scale bar 100 μm; (C) tricolpate striate pollen from stamen, scale bar 5 μm; (D) pair of basally fused carpels, scale bar 100 μm; (E) carpel with stigmatic surfaces, scale bar 100 μm; (F) diagrams of immature flower, stamen and carpel, with scale; (G) floral and inflorescence diagram, with one terminal pistillate flower and two lateral staminate flowers (from Drinnan *et al*. 1991.)

A

B

Fig. 10.8. Reconstructions of *Caloda delevoryana* from Cenomanian rocks of Kansas. (A) Reconstruction of axis before abcission, ×2; (B) detail of one secondary axis with carpel cluster, ×10 (from Dilcher and Kovach 1986).

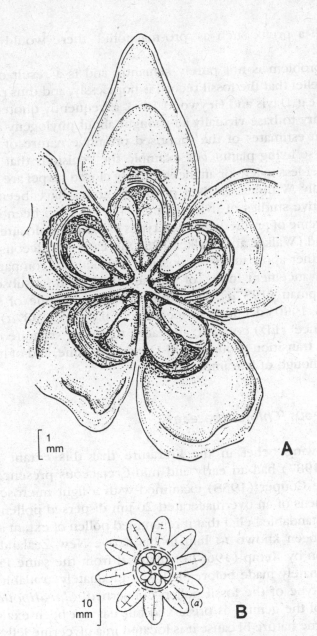

Fig. 10.9. Actinomorphic flowers from the Mid-Cenomanian Peruc Formation of Bohemia. (A) *Asterocelastrus cretacea*, pentamerous locules and calyx, with two anatropous seeds per locule, (B), *Kalinaia decapetala*, scale indicated (from Krassilov and Pacltova 1989, Velenovsky and Viniklar 1926–31).

term 'magnoliid', although strictly it should only relate to the Subclass Magnoliidae. There is no precision possible when anything from one to eight or more families may be involved in such a general attribution, without any statement of characters. Consequently such terms are here avoided, although if they were regularly written with a prefix such as 'pro-magnoliid' there would be less objection.

This terminologic problem is not purely semantic and is a result of many decades of botanical belief that the fossil record is hopelessly, and thus permanently, inadequate (see e.g. Davis and Heywood 1963, a frequently quoted text); it has become customary to base virtually all thought about phylogeny and the course of evolution on estimates of the supposed primitive nature of certain plant characters of these living plants. For example the decisions that certain characters such as vessel-less xylem or an incompletely closed carpel are primitive, and that the Family Winteraceae is a primitive group have been made entirely from comparative studies of living plants. This has even been carried to the unprofitable extreme of predicting which types of fossil should after more diligent search be found (Walker and Walker 1984) in early Cretaceous strata. Although other and further approaches are possible, this kind of comparison is summarised by the 'Advancement Index' of Sporne (1980) and its subsequent rearrangement by Chapman (1987a). Realising the limited scope of such a database, Sporne (1974, 1980) declined to speculate. Chapman (1987a), using her 'Evolutionary Distance' (ED) concept, appeared to favour separate origins of hamamelids as a core transition-group in mid-Cretaceous time, and of magnoliids as independent although of less importance.

10.7   Early Cretaceous 'Chloranthaceae'

There has been a growing belief in the literature that this extant family (reviewed by Endress 1987) had an early and mid-Cretaceous presence; this has persisted ever since Couper (1958) examined with a light microscope a small number of specimens of an overmacerated 20 µm dispersed pollen grain and recorded his understandable belief that it resembled pollen of extant *Ascarina*, which genus had been known to him in his native New Zealand. The follow-up re-examination by Kemp (1968) of material from the same English source was also unfortunately made before SEM was adequately available. The Cretaceous–Aptian holotype of the fossil pollen concerned, *Clavatipollenites hughesii* (type-species of the genus), is preserved but has not been examined by SEM to ascertain its true nature because it is located in a glycerine jelly slide (made by Couper in about 1955) that also unfortunately contains adjacent holotypes of several other palynomorphs. The relevant sample (KCE 475) has been reprepared and the newer preparation has been re-examined using SEM, revealing ten or more species-level taxa of fossil MCT pollen (Phase 4; see Chapter 9) to any of which the original holotype may have belonged. These uncertainties have been recorded in print by Hughes *et al.* (1979) and subsequently, but nothing appears to deflect other authors from a persistent widespread use of the name *Clavatipollenites* and an apparently inevitable attribution to the Chloranthaceae. Muller (1981) maintained that this was the only pre-Cenomanian pollen record of a living family, but his unusually long explanation (Muller 1981, pp. 9–12) illustrated well his own doubts and caution; Chapman (1987a,b) considering both 'Evolutionary Distance' (see Section 10.6 above), and fossil pollen, concluded that 'the choice of Chloranthaceae as an early family was not very appropriate'. This appears to be a clear case for ignoring the original fossil species and genus as uninterpretable, and thus also the many attributions involved with the name.

Erection of a neutral Aptian (Period Classification 120) name for a group of such fossils would be unexceptionable, but it seems preferable to abandon the whole attribution until more convincing mesofossil evidence becomes available. The fossils described as chloranthaceous by Crane et al. (1989) are almost all of much later Santonian–Campanian age and represent a different subsequent problem; the single broken specimen of a fossil androecium from the Late Albian West Brothers (Maryland) locality, which Crane et al. (1989) unfortunately considered to be chloranthoid but at the same time not quite worthy of a name to make a proper record, does not alter my conclusions above, nor does their protracted argument (Crane et al. 1989, pp. 218–24) in support of their speculation.

10.8   Mid-Cretaceous paleoenvironments

Accepting Cretaceous Cenomanian and probably Albian angiospermous fossils with credible but very small and inconspicuous flowers, it is appropriate to search for possible contemporaneous related biological and geologic events beyond the record of this group. Changes in the nature and distribution of other plant groups (Crane 1987) appear to have been gradual and complementary when considered on such a coarse scale.

Changes among large vertebrates appear to have produced new types of grazing Ornithischia (see e.g. Coe et al. 1987) well before Cretaceous Albian time and not synchronised with these angiosperm appearances; evidence concerning small vertebrates is so far inadequate even to detect any involvement in seed dispersal.

Among insects, Coleoptera (beetles) were probably already much earlier concerned in pollen transfer at least with the Bennettitales (Crepet and Friis 1987), having been attracted to eat the plentiful pollen itself; the attractant to the normally separate female structure is not known and was not necessarily a specialised one at all at this early stage. Angiosperms may have produced nectar but there was little development of perianth in the earliest forms, which probably remained in green/white/ultraviolet 'colours'. Apioids (bees) and other aculeate Hymenoptera, and haustorial Lepidoptera (butterflies and moths) did not become widespread nor in some cases had even appeared until well into late Cretaceous time; no special mid-Cretaceous developments of Diptera are suspected. Perhaps the mid-Cretaceous changes in the Pollenifera to produce angiosperms were not so momentous at the time as they have been presumed to be by human observers.

Such was probably the case also with geologic phenomena: the earth was warmer than now but this was probably expressed in a gentler latitudinal temperature gradient (but see Frakes *et al.* 1992, Spicer and Corfield 1992), the sea-level was said to be higher but locally confused by isostatic movement, some soils had become elaborate as in the earliest Cretaceous rendzinas (Francis 1986) but there is insufficient description yet (Retallack 1986) to track the development of the more normal straightforward non-calcareous cases. There is a strong suggestion from palynology and sedimentary facies that early angiosperms evolved not far from sea-margins (Doyle 1978*b*), but overall there is much more paleoenvironmental investigation needed of the rapidly accumulating paleobotanical finds. It is premature to expend argument on the speculative 'ecological-adaptist' evolutionary model of Doyle and Hickey (1976), which itself hangs on the non-fossil theories of Stebbins (1965, 1974).

10.9   Discoveries highlight more problems

The megafossil and mesofossil descriptions of the last decade have by their success underlined the disparity between the information derived from fossil leaves, flowers, pollen, wood, and their paleolatitude distribution. In particular there is a dearth of well-dated facts from the 10–12 million years from mid-Aptian to mid-Albian time, which includes the elusive Potomac Zone IIA. Leaf laminae with reticulate venation were simple (see Fig. 10.3) and very rare, being still known in this interval only from one or two localities. Flowers so far discovered, with few exceptions such as the isolated example from Koonwarra in Australia (Taylor and Hickey 1990), were yet to come from Late Aptian to Late Albian time. Pollen, as for example in the little described Potomac Zone IIA, appears to be placed between a declining, or no longer expanding, MCT assemblage of monosulcates and a very slowly developing triaperturate presence. The latter was probably derived from the former but no individual dispersed pollen so far discovered serves to encourage such a view (Herngreen and Chlonova 1981, Penny 1991). The attractive idea of low paleolatitude origin and successive migration polewards recently favoured by Penny (1992) has languished because of uncertain stratigraphic datings from Africa; it may have all occurred so quickly as to be below current stratigraphic resolution especially in view of the work of Trevisan (1988, 1994). Even fossil wood with vessels appears not to pre-date Cenomanian time, possibly excepting *Sladenioxylon africanum* of Giraud *et al.* (1992) if the stratigraphy can be clarified, and still excepting the 'Aptian' woods of Stopes (1915), which I regard (Hughes 1976) as suspended until further discovery is made at the one coastal locality still available for collecting; despite numerous observers nothing has been recorded in this century.

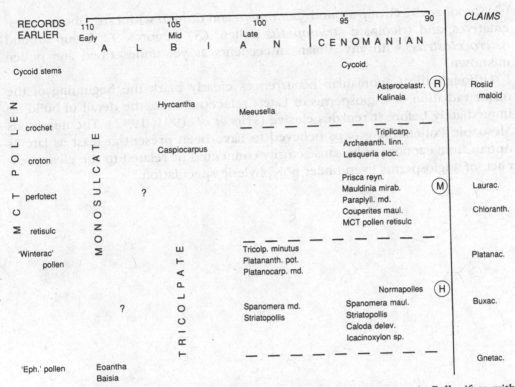

Fig. 10.10. The Albian and Cenomanian occurrences of Mesozoic Pollenifera with angiospermous features. Time-scale in millions of years ago. Claims (right-hand column) are of extant families believed by some to be involved. Some names of families, genera and species have been abbreviated, but full names are in the text. Cycoid., *Cycadeoidea*; 'Winterac.', 'Winteraceae'; 'Eph,', *Ephedripites*; Tricolp. min., *Tricolpites minutus*; Plantananth. pot., *Platananthus potomacensis*; Platanocarp. md., *Platanocarpus marylandicus*; Spanomera md., *Spanomera marylandensis*; Asterocelastr., *Asterocelastrus*; Triplicarp., *Triplicarpus*; Archaeanth. linn., *Archaeanthus linnenbergeri*; Lesqueria eloc., *Lesqueria elocata*; Prisca reyn., *Prisca reynoldsii*; Mauldinia mirab., *Mauldinia mirabilis*; Paraphyll. md., *Paraphyllanthoxylon marylandense*; Couperites maul., *Couperites mauldinensis*; Spanomera maul., *Spanomera mauldinensis*; Caloda delev., *Caloda delevoryana*; Laurac., Lauraceae; Chloranth., Chloranthaceae; Platanac., Platanaceae; Buxac., Buxaceae; Gnetac., Gnetaceae.

The Late Albian–Cenomanian angiospermous reproductive structures are clearly divisible into three unrelated groups, even without actually expressing affinity to extinct groups or families. The neutral nomenclature employed here for these groups (Fig. 10.10) is intended (a) to include the age of the Period Classification concerned (Cenomanian at 90 Ma), (b) to accept a link with the current fashion in affinity speculation, and (c) to allow for unambiguous updating to accommodate new fossil discoveries and reinterpretations (e.g. for Cenoham 93 Group, and so on). The three groups now are adequately expressed as: (1) Cenomag 92 Group around *Archaeanthus*, *Mauldinia* and *Prisca* with probable Late Albian representatives, but associated pollen uncertain;

(2) Cenoham 92 Group around *Spanomera* and *Caloda* with Late Albian representatives, and tricolpate *Striatopollis* pollen; (3) Cenoros 92 Group around *Asterocelastrus*, with any Albian antecedents as yet undescribed and pollen unknown.

Although the Cenomanian occurrences clearly mark the beginning of the major radiation of angiosperms of Late Cretaceous time, the detail of build-up immediately before it remains elusive (Friis *et al*. 1991, 1992). The numerous Mesozoic Pollenifera groups believed to have been present at least as late as Aptian time cannot yet be satisfactorily connected or related to the clear presence of angiosperms even under polyphyletic speculation.

11 Cretaceous angiosperm consolidation

11.1 Late Cretaceous paleoenvironment

As the land floras of the world became very different in late Cretaceous time, there were presumably physical changes in parallel. Although the maximum 'anti-glacial' development between the Permian glaciation and the current Pliocene–Pleistocene–Holocene glaciation fell in Late Cretaceous time, the turning point has been variously placed from Campanian (Singh 1988) or Coniacian (much marine evidence) or even from Albian time onwards. Despite confident but somewhat circularly argued assertions of temperature falls on land, the maximum of anti-glacial effect appears to fit observed facts of land plant radiation most easily in Campanian and perhaps Maastrichtian time, after which the sharpest general changes on land have been recorded.

The effect of a (slightly) warmer earth appears to have been registered in more gradual paleolatitude change to presumably warmer poles (Spicer and Chapman 1990); consequently estimates of climate change and reversal need to be broadly based (Wolfe 1987, Wolfe and Upchurch 1987) on a range of sampling in different paleolatitudes, which has not yet been possible for Late Cretaceous time on the scale of the various studies by Wolfe (e.g. 1978, 1985) for the Tertiary. Another difficulty is that evidence for vegetational development and for elaborate paleosols beyond the land area of the traditionally plant-occupied aggradational land is still lacking. Rainforests (see Wing and Tiffney 1987) and savannahs appear to be environments of Cenozoic origination; this also appears to apply even to general deciduousness of trees (Wolfe 1987) at least in lower latitudes, so that any interpretation of the Cretaceous environment directly from Holocene experience remains very uncertain and difficult. Global coal formation, which had been high in Carboniferous and Permian time, was very low throughout the Triassic and Jurassic periods (most deposits were in Asia) and rose markedly in the late Cretaceous in China and then in Tertiary time, mostly in North America.

11.2 Late Cretaceous 'dicotyledonous' fossils

There was clearly a radiation of plants with net-veined leaves from Cenomanian time onwards, although the scale of the further development in Maastrichtian and Early Cenozoic time was much greater (see Muller 1981, p. 116). The radiation may be perceived quite differently through studies of fossil leaves, fossil flowers, fossil pollen, fossil wood, or the supposed Cretaceous existence of living plant families; only the first and the last of these have yet received much organised attention in the literature.

11.3   Fossil leaves

The majority of the generic names given to these Cenomanian through Maastrichtian fossil leaves incorporate their author's belief about affinity among extant groups. It is as a result difficult to dissociate such suggestions subsequently when necessary from the fossils observed; it is further very difficult to express any grade of uncertainty or doubt (Spicer 1986) against such a loaded generic name armed with nomenclature priority in print. The specific epithets concerned have mostly been constructed from topographic names and form the main reference to the actual record. The generic names in use, however, may have the advantage of providing from their general range a quick overview of the type of foliage that existed at any one time and place (as described by Vakhrameev (1991) for Eurasia), despite the cost of severe limitations in recording and in indexing and retrieval of basic data.

The North American continent from 30° to 70° N paleolatitude has many floras composed and discussed by Wolfe and Upchurch (1987), who, however, expressed their view that a majority of the leaf floras involved require a complete modern updating. Upchurch and Wolfe (1987) employed collective floral characters rather than taxa in their interpretations, in part as an answer to the difficulties they mentioned (see Greenwood 1992). In the middle paleolatitudes of Eurasia, the Euro-Sinian province described by Vakhrameev (1991), the most prominent leaves (Crane 1987) are of plants that have been loosely called 'hamamelids' or even 'higher hamamelids' under such names as *Platanus, Credneria* and *Debeya*. It seems preferable to accept that they undoubtedly have this resemblance (see Crepet *et al.* 1992), while one still seeks to record such single organ fossils in taxa neutral to any supposed platanoid association. It is logical to regard these as probable descendants of the Cenoham 92 group (see Chapter 10), while recording that fossil successors to the Cenomag 92 and Cenoros 92 groups are far less obvious until Tertiary time. It is only in Maastrichtian time that the general diversity of leaf and of other fossils becomes sharply greater.

11.4   Late Cretaceous palynomorphs

The Normapolles province of dispersed palynomorphs corresponds with the middle northern paleolatitudes, although some similar parallel types are now also known from the Austral Region (Jarzen and Dettmann 1992). Following the work in Sweden of Friis and Skarby (1982) and many subsequent papers, some Normapolles pollen is now known to have been produced by flowers of 'hamamelid' plants in Santonian–Campanian time. The many other types of dispersed pollen known in this and in the more northerly *Aquilapollenites* province have been less securely associated with other plant fossils, and the very careful summary of Muller (1981, 1984) lists numerous extant families with which to compare them. Much exploration, however, remains to be done. A peculiarity is the disappearance by Turonian–Coniacian time even in Laurasia of virtually all the pollen of *Ephedripites* and types assumed to be related (grouped by Crane (1987) as ephedroids), and of almost all traces of cycad-like plants; the few extant genera of the latter group can only count as negligible in world floras of today and may thus be considered as having never recovered.

11.5   Maastrichtian floral radiation

Crane (1987) and Wolfe (1987) recorded a great increase of angiosperm diversity in a much enhanced representation of living families. Muller (1984), estimating the presence of angiosperm orders from pollen records, showed a greater increase in Maastrichtian and Paleocene time than in any earlier or subsequent period. This radiation also appeared to introduce and include the largest and most striking plants of 'monocotyledonous' type.

11.6   'Monocotyledonous' fossils

Leaving out of the account for the moment such early fossils as the Triassic Carnian *Sanmiguelia* (see Section 8.11), which has no connecting links over the succeeding 100 million years, the earliest serious claims are from Cretaceous Aptian rocks of the eastern United States and elsewhere (Doyle 1973). Small reticulate monosulcate columellate pollen with pattern-variable size of lumina, included in *Retimonocolpites* by Doyle but resembling Late Cretaceous *Liliacidites* of Couper (1953), are present in numerous assemblages (Fig. 11.1); they are recognised by Muller (1981) as being of a general type found in several extant monocotyledonous families but not closely identifiable with any one of them. After these occurrences there is then a long gap because Muller (1981) accepted no further records of pollen of any Liliopsida (Monocotyledons) until the palm, pandan and Restionaceae pollen that appear in Maastrichtian strata over 50 million years later. The Aptian records are universally part of the variety of the problematic small grain monosulcate (MCT) assemblages (see Chapter 9) and their ultimate affinity will probably be settled with further study of that group of fossils. Apart from a few indeterminate parallel-veined leaves, Cretaceous megafossils begin with some unusually fine-grained palm wood *Palmoxylon cliffwoodense* (Fig. 11.2) from the Coniacian of the Magothy Formation of

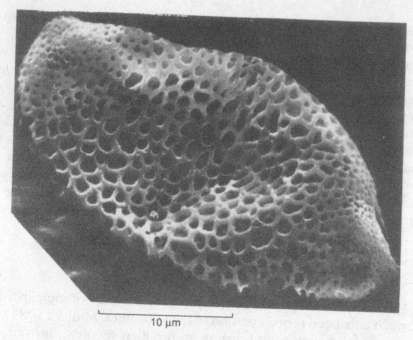

Fig. 11.1. SEMG of pollen *Retimonocolpites* from Potomac Formation Zone IIB from Delaware City Borehole D13 at 545 feet depth (from Doyle 1973).

New Jersey in paleolatitude 35° N (Berry 1916). Numerous subsequent Late Cretaceous palm wood records are approximately from Tethyan shore regions between the paleoequator and 40° N, from Kutch, to France and Germany, to Mexico; in contrast Maastrichtian dispersed 'palm' pollen has a more South Atlantic distribution (Herngreen and Chlonova 1981).

11.7   Existing theory of origin of 'monocotyledons'

Taking the first actual fossil specimens as the Coniacian palm wood referred to above (Fig. 11.2), and relegating the Carnian *Sanmiguelia* and the Aptian palynomorph fossils as arguably of other and perhaps unknown provenance, the difficulties created by patterns of theory become apparent. For several decades, the putative first angiosperms have been 'dicotyledonous' with 'monocots' descended from them (see Walker and Walker 1984); although direct fossil evidence for this was lacking, the theory only caused difficulty when imagination was also precluded. Burger (1981*a*) chose to revive an earlier alternative 'monocot' theory for the main angiosperm origin, with the 'dicots' descended later. This drew even more heavily on imagination than did the fossils, and also suffered from the same grievous limitation of assumed monophyly for all angiosperms. In presenting his argument Burger (1981*a*, table 1, p. 211) carefully stressed the large number of detailed differences from chemistry through to gross morphology that characterise 'monocots'. Whereas other characters of

Fig. 11.2. Petrified specimen of the earliest known palm wood, *Palmoxylon cliffwoodense* Berry from the Magothy Formation (Coniacian) of New Jersey. 1, transverse cut, ×1; 2, the same, ×2; 3, longitudinal cut, ×1; 4, photomicrograph of transverse section, ×20; preservation in limonite (from Berry 1916).

'dicots' were widely parallelled in gymnosperms, those particular to 'monocots' were more often shared with the 'more primitive' pteridophytes. The characters of 'monocots' embrace scattered vascular bundles in the stem, very little branching, simple undifferentiated leaves with lightly developed cross-venation and sheathing leaf bases, an embryo with one cotyledon because the leaves are fundamentally alternate, an endodermis throughout the root and also the stem in most cases, vessels of a type seen (though rarely) in pteridophytes, starch grains (in seeds) nearer to pteridophytes than to 'dicots', and types of lignin found in all land plants other than dicots. Consequently the 'monocot' origin for all angiosperms (including dicots) sounds no more attractive than does the standard dicot-first origin in a so-called woody 'Ranalean' framework.

11.8   Suggested separate origin for palms

Liliopsida (monocots) appear to be so different from other plants, not only in the many characters referred to (see Section 11.7 above) by Burger (1981a), but also in the difficulties encountered by Lowe (1961), Sporne (1974, 1982) and Chapman (1987a) in establishing characters for an 'Advancement Index' for them. If 'monocots' were merely trimerous variants on a normally pentamerous 'dicot' flower theme from a single united origin, why should there be so much difficulty in comparing characters?

In answer to this difficulty a more likely solution appears to lie in an entirely separate origin of palms from such pteridophytes as the eusporangiates or other ferns without any 'gymnosperm' connection at all. Although this runs counter to most botanical tradition, it seems at least worthy of close study for a few years as part of a postulated fully polyphyletic origin for all angiosperms.

It also happens that, with palm stems as the earliest agreed 'monocot' fossils, it is relevant to record that the only other major group of land plants that have attempted any such unusual false-trunk tree architecture as is seen in palms have been the tree-ferns of various kinds from Carboniferous time onwards. These tree-ferns appear to be represesented in Cretaceous Albian time by the unexpected and still enigmatic *Tempskya* group (Ash and Read 1976) mentioned in Section 8.35. Even in the nineteenth century this suggested parallel of palms with *Tempskya* might not have caused much surprise because the genus *Endogenites*, which preceded *Tempskya* for the Cretaceous Valanginian *T. schimperi* from the Lower Wealden rocks of England, was originally erected as a genus for Tertiary fossil palms by Brongniart (1822). Seward (1894, pp. 148–58) provided a long, laboured explanation of the probable but unproven nature of these fossils as fern stems, although by then they were no longer thought of as palms.

11.9   Ferns radiated strongly in Early Cretaceous time

Early pteridophytes appear now to have been originally restricted in habitat to at least seasonally moist situations by their vulnerable separate gametophyte generation. Natural survival (or selection) has subsequently led some pteridophytes to unexpected existence in other styles and habitats. *Lepidocarpon,*

Mazocarpon and *Miadesmia* in Carboniferous time reached a seed-habit in all but name, although this was not apparently successful beyond that period, possibly because of the nature of the whole plant. False-trunk tree architecture was employed by *Psaronius* and *Thamnopteris* in Carboniferous and Permian time and by *Osmundites* separately later. These developments were not apparently accompanied by any striking change in sporophyte reproduction and therefore have never called for special comment because *Cyathea* (and allies such as *Hemitelia*) and *Dicksonia* live in this style today, albeit in restricted upland tropical and southern temperate environments, respectively.

In Late Jurassic–Early Cretaceous time there appears to have been a strong pteridophyte (fern) radiation into both perpetually humid and then aquatic life niches in which the standard alternation of generations required special modifications. This radiation is recorded in megaspore developments progressing towards but not reaching a seed-habit and sometimes accredited to Cretaceous 'Marsiliales' and later also to truly aquatic 'Salviniales'. Independently, in another direction well above ground in possibly xerophytic surroundings, there appeared the Cretaceous tree-forms of *Tempskya*, which has long been assumed to be another tree-fern and is normally referred to as such. Andrews and Kern (1947), however, regarded *Tempskya* as uniquely different. They could only speculate that the still unseen vegetative parts must have been numerous and very small because the vascular strands of the trunk were so small; above a trunk base entirely of rootlets there were many vascular strands which they regarded as independent stems rather than steles. The reproductive structures were also still completely unknown to them and even much later to Ash and Read (1976), who reviewed the North American members of the group. In the same geologic period, ferns usually recorded as 'schizeaceous' radiated strongly (*Cicatricosisporites, Trilobosporites*), probably to new environments, although this is based almost entirely on dispersed palynomorph data (for caution on attributions, see Dettmann and Clifford 1991, 1992). Shortly afterwards the foundations were also laid for the modern dominant Polypodiaceae, which ultimately in Late Cenozoic time produced a strong tendency to epiphytic habit in life under closed tree canopies.

All of these developments were far removed from the straight-forward Jurassic ferns *Coniopteris, Cladophlebis* and *Gleichenites*, and illustrate the widespread significance of this fern radiation into Cretaceous anti-glacial conditions, a phenomenon that has been masked by the greater size and variety of plants of the Pollenifera radiation occurring above them, and by the much emphasised botanical separation of seed-plants from pteridophytes.

11.10 Palm fossils

Permineralised palm wood fossils, being relatively indestructible, leave many records from Coniacian time onwards, but other fossil palm remains such as leaf compressions are less common. Distinctive vegetative fronds are recorded as fossils from Maastrichtian time but the few earlier claims from Cenomanian time onwards are not convincing (Daghlian 1981); such remains are well known and undisputed in the Tertiary (see e.g. Erwin and Stockey 1991). Palm pollen (Muller 1981) is also clearly present from Maastrichtian time onwards. Reproductive structures have not been identified until much later in Tertiary time (Knobloch and Mai 1986); inflorescences of living palms show a most unusual variety of characters of morphology and of development (Tomlinson 1961,

Fig. 11.3. Characteristic stem and aerial roots of *Pandanus* tree, growing (1992) in a garden in Durban, South Africa.

Corner 1966), and consequently provide almost no guidance concerning the possible reproductive pattern in the Cretaceous palms.

The considerable variety of over 2500 species of living tree-palms, ground-palms and rattans of tropical and subtropical maritime regions, and the differences in the Old and New Worlds are both a warning and an encouragement to ultimate understanding of Tertiary paleobiogeography. Living palms are less closely investigated and classified than most other plant groups because, on account of their size and form, they are not easily subjected to herbarium study. Although palm wood and large parallel-veined leaves make durable and distinctive fossils and are thus likely to provide a particularly positive record, it seems possible to go along with the belief of Corner (1966, and oral communication 1992) that tree palms were the first monocotyledons (with pandans) and that rattans and ground palms and all monocotyledon herbaceous families were derived from these at later dates. Thus although it may be possible to derive all other monocotyledon families from palms (in the broadest sense) in Maastrichtian time onwards, they could also be of separate origin if the pollen with variable reticulate exines from the Aptian monosulcate tectates (see Chapter 9) be accepted. Although such evolution of the palm-growth 'cabbage' appears botanically unlikely at present, *Tempskya* as some kind of proto-palm could still even prove to be a seed-plant itself rather than a tree-fern.

11.11   Pandan group fossils

Leaves attributed to this group have been reported from Maastrichtian rocks particularly of Romania (Vakhrameev 1991). Dispersed pollen of pandans of Maastrichtian age from North America was accepted by Muller (1981) as the first record of this group (Jarzen 1978). Other later records under the name *Pandaniidites* have also been made from the Maastrichtian of North America, and throughout the world in Cenozoic time, mostly under the extant generic name *Pandanus*. Living, there are over 600 species in coastal or marshy areas of the Old World tropics; they are small trees and shrubs known as screw-pines and some bear cone-like fruits slightly resembling pineapples; the leaves are often twisted and large aerial roots are prominent (Fig. 11.3). The small New World tropical group Cyclanthaceae also show some resemblance, but neither appears to be close to other monocotyledons. The argument for possible pteridophytic origin applies here (as with palms), and the choice of ancestor would fall in this case on the group of surviving lycophytes such as *Isoetes* relatives and Cretaceous *Nathorstiana*, perhaps descended from Triassic *Pleuromeia*. This is at present pure speculation, but worth some study and testing as a separate origin for a sizeable and prominent living group of plants that appear to be isolated even among monocotyledons.

11.12   *Late Cretaceous general comment*

Most of our current knowledge comes from northern middle paleolatitudes; very little apart from some dispersed pollen is known so far from equatorial regions (Muller *et al.* 1987, Salami 1985, 1990). The purpose here is not in any sense to offer systematic description, but to comment on trends and on apparent anomalies, the most prominent of which is the separate and rather late appearance and radiation of 'monocotyledonous' plants. This can be seen as palms and pandans relatively quickly giving rise in Maastrichtian time or shortly after to some of the herbaceous families. Alternatively the Liliaceae and allies may be considered to have arisen separately and much earlier, using the records of Doyle (1973) of Aptian monosulcate pollen; subsequent supporting evidence has not been found and therefore this solution appears to be unnecessarily complex unless the Liliaceae group are indeed of separate origin altogether.

12 Cenozoic angiosperm radiation

12.1 *Paleobotanical indigestion*

There have never yet been enough paleobotanists in the world to study all the innumerable individual Tertiary plant fossils in a meaningfully constructive global manner, in addition to the more customary work on the more remote fossils of earlier periods. The greatest ever radiation of organisms, primarily land plants and insects, took effect in early Cenozoic time (65–50 Ma) and apparently continues now unabated. From the then-living Paleogene plants, abundant leaves, wood, flowers, seeds and palynomorphs have been recorded, including some excellent random preservations in amber, lignite and permineralisation. As with all plant fossils, the specimens invariably consist only of individual separated organs for which it is consequently very difficult to establish connections with any other fossils; this is currently the case with leaves, with or without cuticle, and palynomorphs. Only a very careful recording and statistical treatment of occurrence association data will provide any solution to this problem. As recorded in Chapter 1, the main paleobotanical textbooks consequently still tend to treat the whole topic of Cenozoic plants as unsatisfactory and avoid it.

12.2 *Climate patterns and angiosperm radiation*

The timing of this great radiation phenomenon appears to be most simply explained in terms of evolving climate pattern (Wolfe 1985) interacting with the plate tectonic geography. It rides on the back of a world climatic pattern change more or less constantly in one direction from the Late Cretaceous anti-glacial (a warm period that endured perhaps 30 Ma but included fluctuations) through to the present almost certainly unfinished glacial period; this currently shows the latest of a number of minor reversal phases, and may well in this world as a whole be deemed to have begun long before the traditional Quaternary period opened (Frakes *et al.* 1992). The main biological product has been what are now called tropical conditions exemplified by the current species-rich tropical rainforest, which may be viewed as climax or interim climax of a kind with no parallel in geologic time since the early Permian, when the continents were entirely differently placed and assembled.

12.3   The Cretaceous/Tertiary boundary phenomena

The recently reconstructed and much discussed episode named after this K/T boundary was clearly only dramatic in a closely selective manner. Taken over as much of the world as has yielded fossils of relevant age, the great angiosperm diversity of the land areas of Maastrichtian time moved smoothly on to become the even richer Paleocene and Eocene floras. These showed evolutionary changes but without any sign of the universal catastrophic devastation suggested by the more enthusiastic geologists and others studying impact phenomena (e.g. McLaren and Goodfellow 1990, and many more). Undoubtedly the dinosaurs, some cephalopods and some planktonic foraminifera suffered extinction or major evolutionary change near the K/T boundary, but numerous other types of organism represented by fossils, both on land and in the sea, did not. The well-publicised 'fern-spikes' suggesting palynologic evidence for floral recovery from disaster may well have been a North American regional effect, perhaps from fire or even impact driven, but not universal.

The K/T boundary was obviously marked by some important progressions in the organic world such as the change to leaf deciduousness (Wolfe 1987), which could also be attributed to a downturn from a climatic maximum of anti-glacial nature. The temporary absence of certain animals, both herbivores and their predators, may even have given impetus to some further aspects of early Cenozoic plant life as part of the evolutionary pattern. Both the general continuity and the maximum of creation of new plant taxa testify against any world-wide destruction. Impacts were real enough but had little effect on global organic evolution.

12.4   Eocene floras

The famous Early Eocene floras of the London Clay and adjacent strata (Reid and Chandler 1933, Chandler 1964, 1978, Collinson 1983), and those more recently described from Messel-bei-Darmstadt in western Germany (Collinson 1982), present a suggestion of apparently tropical environments in paleolatitude 40° Eocene N, which has always been difficult to accept in such a latitude (see treatment by Wolfe 1985). The 400 mile proximity of the London Clay environments to roughly contemporaneous floras in Scotland with no plants in common has increased the difficulty. The undoubted affinity to living families of numerous London Clay plants, prominently led by fruits of *Nipadites*, resembling those of the stemless *Nipa* palm of current mangrove habit, need not be challenged; the often brilliant interpretations by Chandler and others from the exacting pyritic preservation will endure. It has become apparent that, although *Nipa* is confined now to truly tropical environments, the fossil *Nipadites* and other associated plants and their many familiar characters were existing in merely warm temperate (but frost-free) conditions in Eocene Britain. There were no tropical rainforest environments at that time even in much lower

paleolatitudes, because such environments did not evolve until full glacial circumstances induced them by increasing the latitudinal temperature gradient in
the tropics of Miocene to Pleistocene time.

The great diversity of taxa recorded from the London Clay and parallel occurrences further east was gradually squeezed southwards from 40° N paleolatitudes towards Tethys through Late Eocene, Oligocene and Miocene times, by
the overall pre-glacial cooling expressed in these sharper latitude temperature
gradients. It so happened in Cenozoic East and South-east Asia that the northern
land area extended far enough south into equatorial regions throughout this
period and hence these plant families, now called tropical, were able to migrate
south with overall falling temperature and to survive there and evolve further.
The descriptions by Chandler (1964 and earlier *in litt.*) of successive late
Eocene floras from the Bournemouth Beds up to Headon Beds (southern
England) illustrate the nature and rapidity of this progressive change in which
many members of Early Eocene floras disappeared out of southern England in
2 to 3 million years (Collinson and Hooker 1987). Because of the accidents of
plate tectonic geography, East Asia from 60° N down to the equator remains
the most continuous and instructive area of the world for plant migrations
related to angiosperm radiation through Cenozoic time; in the European sector
Tethys was unfortunately too wide and extended too far north to permit such
southward migrations. Collinson and Hooker (1987, p. 268) suggested that
the 'tropicality' of the London Clay flora may have been underemphasised for
taphonomic reasons, but the implications of the better term 'tropical potential',
which they use elsewhere, appear to indicate a more satisfactory terminology
and explanation.

The question of diaspore size has recently been discussed (Tiffney 1984,
Wing and Tiffney 1987): recorded Cretaceous diaspores were small; in contrast
the largest extant diaspores come from tropical rainforests where many huge
seeds with easily observed large endosperm resources are used to propel seed-
lings rapidly upwards towards the light from the low illumination of the forest
floor. Tiffney (1984) concluded that Tertiary floras such as the London Clay
Early Eocene (Collinson 1983) and the Clarno of Oregon Late Eocene (Scott
1954) showed fairly large average size but that this was boosted by small numbers of very large seeds in a general representation of rather small seed size.
The typical sizes in the London Clay (Collinson 1983) appear to be very much
smaller than those of comparable living plants in extant tropical forests of Malay-
sia, perhaps supporting the theory of the migration pattern postulated above.

12.5   Taxonomy, nomenclature and evolution

One of the chief difficulties in handling the enormous quantities of Cenozoic
data has been automatic adherence to nomenclature-bound taxonomy. As with
evolution during the earlier geologic periods, the main problem is the piecing

together of successions of events as far as this is possible from determining associations of all kinds of available fossils, which are (including palynomorphs) very numerous and particularly diverse throughout the Cenozoic world. Because the Cenozoic is relatively nearer than the Mesozoic to the present day in geologic time, the temptation is strong to identify fossils in some way with living plants and thus to 'understand them better'. The chance of finding bio-chemical evidence is much greater in the Cenozoic than in the Mesozoic and therefrom perhaps the possibility of constructing more (non-fossil) phylogenies. Ironically in the Cenozoic the danger to the satisfactory unravelling of evolution through fossil occurrences comes not from the influence of genetic and nuclear paleontology or from fossil-weak cladistics as it does in the Mesozoic, but from the actual study of fossils by well-informed botanical experts who are over-anxious to record comparisons.

As any reader of Chandler, Collinson, Wolfe, Tanai, Boulter, Manum, Scott, Krutzsch, Stockey and very many other authors will recognise, the fossil organs (fruits, seeds, pollen, wood) most closely compared with living plant genera or families for paleoecological purposes almost always include one or more puzz-ling or unexpected characters even in the detail of the one organ available. Fossil leaves additionally present further problems because of the known diver-gence of leaf type in time and/or space in so many living plants, which is pre-sumed to inflict similar uncertainty on the fossils. The 'identifying' of perfectly good organ fossils into genera and families of living plants produces two prin-cipal kinds of waste of information: (a) the obscuring and thus loss of detail of the actual records of occurrence into nomenclature-bound lists of taxa that are merely secondary opinion, however strong the belief and experience behind them; and (b) the production of so-called long-ranging genera and families that if accepted preclude further investigation of their evolution in each case.

It is surprising that these curiously inaccurate taxonomic practices are still applied by many serious workers, and set in concrete by revered nomenclatural priority and other rules. Every single character of a separated fossil organ is potentially needed in the database and must be retrievable where numbers of characters available always tend to be inadequate for any single purpose. A genus of living organisms must be completely reviewed if emendation or addition of species is planned, but this cannot be properly carried out if single-organ fossils are included as whole named taxa. The only purpose of creating necessarily artificial organ taxa of fossils and the naming of these is for the making and retrieval of individual records and the sequencing of these into evolutionary interpretation; the wrongly directed submergence of any fossils in living taxa prevents the proper development of this activity.

What is necessary for the successful recording and emphasising of the differ-ences in fossils into an evolutionary chain of small changes is a slightly different taxonomic philosophy that is perfectly possible despite a generally observed desire to stay with a nomenclature compatible with the Linnean binominals used in working on Holocene floras. The recommended change is to insist that all organ species of fossils (virtually all fossils) be placed in genera employed and defined for fossils alone. Families on the other hand, are solely concerned in interpretation, are not normally indexed for retrieval, and can therefore be ignored as not affecting the database.

The simplest change in this direction would be to employ the original (and additional) Lyellian stratigraphic prefixes of Paleo-, Eo-, Oligo-, Mio- and Plio- as appropriate to all generic names of extant organisms employed for fossils e.g. Eo-nipa and Mio-acer. Such prefixes would be short and need not imply that the stratigraphic level is accurately known, but only that the holotype of the type species concerned comes from known Eocene or Miocene strata. Wolfe and Tanai (1987) acknowledged this problem in a large monograph of leaves and seeds of Cenozoic *Acer*, by employing numerous groups and sections, although they retained the extant genus throughout an illuminating and searching statement. Wolfe and Wehr (1988) in a classification of North American rosaceous foliage, also mixed their use of extant and fossil genera. Preferable, to achieve the aims outlined above, would be a change to the paleotaxon and comparison record system of Hughes (1989), which would be clear of the supposed need to maintain nomenclatural priority, which is a complication entirely unnecessary for fossil records. Any system, however, which preserved all fossil records to be individually retrievable, would suffice. Even an agreement that all paleoenvironment predictions are ephemeral and cannot be allowed to leave 'footprints' in the database would represent progress.

12.6 Avoidance of the use of Quaternary methodology

The successful Quaternary exploration over many decades, especially using palynomorphs, differs from paleopalynology in its dependence on the use of types for comparison drawn entirely from reference collections of material taken from the living Holocene flora. This procedure has been excused by the assumption that evolution over such a short time as the last million years was negligible and could thus safely be left out of account. The very success of this Quaternary work, however, has led to a desire to extend the practice further back into Pliocene and earlier time 'to aid paleoenvironment and other interpretation'; such extension should be energetically resisted as becoming less and less soundly based (in fact, anti-evolutionary) with the increasing period of time under consideration. Additionally, an agreed interface (or understanding) with normal paleontologic (in this case, paleopalynologic) practice is required; if the taxon prefixes or other devices recommended in Section 12.5 are employed through to Pliocene time, using normal fossil holotypes for reference, the present Plio-Pleistocene interface could remain accepted for the change of method without difficulty.

12.7   The Cenozoic contribution

Any solution to the current problem of angiosperm origin will for the present almost certainly be based mainly in Mesozoic rocks and fossils, but is likely later to have a profound effect on views of living angiosperm classification. The various classifications in use at present are all based entirely on comparative morphology of living plants and represent the best convenience now attainable without access to meaningful evolutionary succession; there is little sign that such classifications are likely to make a gentle fit when genuine evolutionary successional facts become available. Uncritical acceptance of hitherto unsuccessful monophyletic patterns for dicotyledons and monocotyledons taken together has dominated the scene so much that any change is already resisted, and will probably continue to be so. The importance of Cenozoic plant studies is that the rich transitional material therein will have to be fully mastered and will eventually become the central concern of paleobotanists. Fortunately the handling of so much data can now be safely envisaged, although success will depend on the quality of input, which will require planning of the kind discussed above. There is a strong argument therefore for redeveloping Tertiary records on an urgent basis, and an encouraging start appears to have been made by Herendeen and Dilcher (1992) for the Leguminosae.

13 Research unattained but possible

13.1 In passing

During the preparation of this book, it became clear that various projected investigations were unattainable at present for time, funding or organisational reasons, and others were new because they were simply consequent on suggestions made above. All of those mentioned relate directly or obliquely to the search for Early Cretaceous monosulcate columellate–tectate (MCT) pollen, and are non-polemical in nature; they can be divided into those related to mesofossils, to palynology, to plant megafosils and to animal and other environmental circumstances.

13.2 Mesofossils

Although the existence of small dispersed submicroscopic fossils such as seeds collected through sediment dispersal and sieving has been well known at least since megaspores have been sought (e.g. Dikstra 1949, 1951, Hughes 1955, 1961), their detailed study and particularly their photography (or illustrations) remained very difficult until SEM became available. So-called fossil tea-leaves (term of T. M. Harris, but already outdated) of this kind were encountered in the Wealden by Oldham (1976) during his study of dispersed cuticles two to three years previous to the publication. The truly spectacular results of Friis and Skarby (1981) from southern Sweden and more recently of Drinnan et al. (1990) from Mauldin Mountain and other localities in northern Maryland, USA, cannot be expected to appear frequently, but throughout all the outcrop sections of the Early Cretaceous non-marine sediments of England and western Europe there are intermittently occurring and patchily suitable beds for exploratory study. Surely the best hope of finding any trace of the actual plants providing the MCT pollen must lie among mesofossils yet to be extracted.

13.3 Palynology

As stated above the SEM strew-search for very small pollen must be continued stratigraphically downwards at least to rocks of Berriasian age and probably since the publication of Cornet and Habib (1992) to the Jurassic Oxfordian;

some such work has begun here at the University of Cambridge but as the results so far are either negative or inscrutable they have been omitted from this work. There remains very much more to do, and presumably it will be directly relatable to the whole problem of angiosperm origin rather than simply to elucidating the MCT pollen. Coupled with this search should be the aim of dividing *Eucommiidites, Classopollis* and *Dicheiropollis* pollen of this period into successive taxa, using characters determined through electron microscopic study.

13.4   Plant megafossils

The stratigraphic ranges in different paleolatitude belts of the plants of the *Caytonia, Leptostrobus* and *Williamsonia* groups need complete reassessment. As pointed out in Chapter 8, virtually all knowledge of the plants involved centres on one or two major discoveries, principally from the Jurassic Bajocian rocks of East Yorkshire included in the Period Classification 160. In each case a clear statement is required of what is also recorded from Period Classifications 140, 120 and perhaps even 100. The material usually consists of leaves and less frequently reproductive structures, but all need to be expressed as graded comparison records to taxa known or newly erected.

In the case of *Williamsonia*, several new species of flower have been recorded very fully by Watson and Sincock (1992) from earliest Cretaceous rocks of England, but even these need to have related to them all usable records from other areas that consist of anything more than a tick on a list. Probably in this case an agreed general scheme of scores for association of the different organs should be drawn up (e.g. cuticle identification, same bed, same formation, same estimated age, in this descending order) on lines well initiated by Anderson and Anderson (1985, p. 85) with their comprehensive 'reliability' formula. More remote is a parallel procedure for association of the leaves with dispersed pollen that would itself have to be subjected first to SEM study for adequate character differentiation. In the case of *Tempskya* (see Chapter 11), bearing in mind the failure so far to associate successfully the permineralisation and compression fossils, a breakthrough in connecting even a leaf with the stem is likely to come only from good lateral thinking into a complex distribution and taphonomic challenge.

13.5   Other supporting information

With such recent and rich finds of insects in the English Wealden (Jarzembowski 1991), a structure to the whole assemblage will very probably be worked out. Smaller reptiles of all kinds probably for the most part lived on insects, but a deliberate study of medium and large Early Cretaceous reptile herbivores is needed independently without any reference to the more obvious carnivores; the very large *Iguanodon* and early hadrosaurs existed clearly enough, but that was obviously not the whole story. Finally, the English non-marine sediments of Valanginian–Barremian age may not present very good examples of identifiable types of fossil soil, but there is widespread evidence of oxidised beds and some more elaborate chemical study may be possible.

13.6   Aptian problems in South Laurasia

Although the various related or unrelated types of MCT pollen urgently need elucidation, there appears additionally to be a palynologic hiatus in the long Aptian age (124.5–112 Ma ago). At Phase 5 and later the diversity of MCT pollen declines, *Ephedripites* declines, and triaperturates have barely begun. In Britain there is rapidly increasing influence of marine transgression but it is relatively local; in the North American Potomac Group records are relatively few from this interval. Significant angiosperm leaf records begin in Albian time. Documentation of palynologic detail in Aptian subdivisions or zones is by no means yet adequate, suggesting that a new study area may be required.

13.7   Other regions

Most investigation so far has been made in South Laurasia and the Africa/South America (ASA) provinces; despite the difficulty of accurate stratigraphic correlation to an 'age' or shorter period as exemplified by Koonwarra in Victoria, Australia, there are many more mid-Cretaceous rocks that have escaped destructive metamorphism and that should contribute to more complete global palynologic coverage.

14 Principles, time and choice

14.1 Fossil record paramount

It is only possible to reconstruct a genuine successional (evolutionary) history for a group of organisms through the fullest use of the fossil record, even to the extent that any other ancillary source of information or theory such as extant comparative morphology should be validated against the record before acceptance. The often disputed degree of completeness of the fossil record refers not to the potential of the record but only to a measure of the total investigatory effort that has already been applied to the materials in question up to any selected moment. Dissatisfaction with what has been achieved is frequently aired, but the complaints usually refer to a desired but so far unavailable part of the possible record. They seldom relate to all the information that could have been assembled and integrated but rarely is so placed together because individual fields of study have become progressively isolated. It will be proper to consider abandoning the study of the fossil record only when no further approach can be found and absolutely no further detail can be extracted; in view of all the continuing progressive developments of collecting and of handling techniques, such a suggestion of despair seems entirely inappropriate and unlikely to arise.

This belief having been stated rather more mildly on a previous occasion (Hughes 1976), the criticism was then made by Doyle (1976) in an exemplary and finely balanced review of my 1976 book that the belief was so 'extremist' that it was likely to damage the geologic botanical co-operation that was being sought in my book. Nothing that has been said or written in the last eighteen years has convinced me that a compromise implied by Doyle in this kind of co-operation is in the interests of desired 'sweetness and light' and can be usefully made. It is not possible to hold a half-belief in the significance of the fossil record in this context, with other sources considered as equally good or possibly better. Purely for the purposes of diplomacy it is understandable if necessary to invoke the principle of 'unripe time' (Cornford 1908) if such a view is sincerely held; the purpose here, however, is direct and is to establish a logical method, of which the ultimate consequences need not be of immediate concern.

14.2 Relevance of Holocene studies to the past

The tantalising shortage of information available from most fossil occurrences could not be overcome without the interpretation stimulus and the general inspiration derived from the quite separate comprehensive study of living organisms and of their mode of life. The development of such studies through ever more detail as in current molecular investigations certainly improves the chances of perceptive interpretation of fossils, but does not alter the complete separation of all Holocene studies from any direct application to the less evolved circumstances of earlier examples of past life. Each selected past time, as exemplified by any one of the Period Classifications, should be taken as independent of all that takes place subsequently (later); its interpretation should be kept distinct from that of all those others, and should be based only on its own facts, however inadequate they may at present appear to be.

14.3   Paleobotanical and botanical baggage

Through the fault of no individual, but as a result of numerous factors understandable with hindsight, both paleobotanists and some botanists have become heavily burdened in their search for evidence of evolution. They have an unavoidably huge, untidy and unmanaged classificatory problem with a system for 250 000 living angiosperm plant species, in which much has been invested, and for which urgent practical reasons have often dictated the pattern. Through perhaps insufficient contact with geology and stratigraphy in this century, an apprehensive attitude to the difficulties of the fossil record has built up into a botanical disdain for fossils that is at the least thoroughly unhelpful and unscientific. There has been a late enthusiasm for certain evolutionary theoretical positions such as neoteny, which were already absorbed or discarded by zoologists a generation earlier. Obsession is the only word for the determined stance on monophyletic origin of angiosperms and of other groups; this has not been improved by recent cladistic studies with their regular circular arguments based on necessarily restricted matrices of data. There has been a persistent vague expectation of some single dazzling success perhaps from a fortunate megafossil discovery and a brilliant interpretative description, or even from a cladist's coup, something that would settle this tedious side-issue and not disturb other botanists. The outcome of retaining such baggage is a predisposition to a rather narrow view of this big problem, hence the comforting prolongation of the 'mystery' aspect in some quarters.

Geologists also have had tunnel vision in relatively recent times, but the advent of plate tectonic studies has beneficially lessened some of this in a dramatic manner. It will be at least as helpful if paleobotanists and botanists can bring themselves to automatic acceptance of the paramountcy of the fossil record in all evolutionary studies (Pant and Kidwai 1972, Pant 1990), and in particular to the dispersed palynologic part of the record. This paleopalynology does not yet have a good data-handling reputation but it has to become an integral part of what is almost separately regarded as megafossil paleobotany; there are still two apparently irreconcilable international organisations (IOP, International Organisation of Palaeobotany, and IFPS, International Federation of Palynological Societies) that both continue to find reasons to organise even in 1992 separate international conferences for 1996. With a proper integration of these two, the angiosperm problem (as an example) would be likely to become a single, normal, successful and freely expanding field of study. Indeed the recent important works of the Friis, Crane, Drinnan and Pedersen partnership (see references under these names) appear to offer such solutions, but much further subtle change of attitude will be needed over the whole field of study of the Mesophyticum (see Section 14.7).

14.4   Time and the angiosperm definition

The clear separation of an angiosperm from other seed-plants can be made with relative ease in Holocene time using Holocene (extant) material. At any earlier time in the history of seed-plants (Pollenifera), the available characters for such a purpose are few and markedly incomplete. A choice of transition time has therefore to be made merely as a matter of agreed convention and, although much remains to be discovered, Cretaceous Late Albian provides both frequent examples of triaperturate pollen and of 'dicotyledon' net-veined leaves. This choice leaves the Barremian–Aptian MCT pollen within the Pollenifera despite the tectate nature of the pollen, but is an advisable choice until a plant or at least an acceptable leaf also becomes known from that age. Such a decision on Late Albian (at 100 Ma ago) still determines nothing precise about development of carpels or of flowers.

14.5   The plant beginnings of the Mesozoic era

The Paleozoic Permian period has produced some plant fossils included in this investigation and is introductory to the Mesozoic era; the floras concerned and their possible relationships have been categorised by Meyen (in Vakhrameev 1991). The generalised distribution of these floras (Fig. 14.1) expresses the end of the late Paleozoic glacial maximum of Early Permian time; it appears that land organisms were constrained principally by physical factors and features rather than by any biological competition caused by entering a climatic pattern that was steadily ameliorating. Much more attention has been given by geologists and paleontologists to the pattern of events in the adjacent shelf seas that was subject to the same overall theme but was complementary, ocean current controlled, and not therefore very obviously in parallel. The very extensive End-Permian shallow marine invertebrate extinctions (not reflected on land) have been explained in terms of tectonic, sedimentary and other changes but these were perhaps aided (or guided) by climatic warming change that was in effect destroying the previous very successful tropical benthonic filter-feeding regime. The essentially predatorial cephalopods nearly failed and disappeared

260 Ma

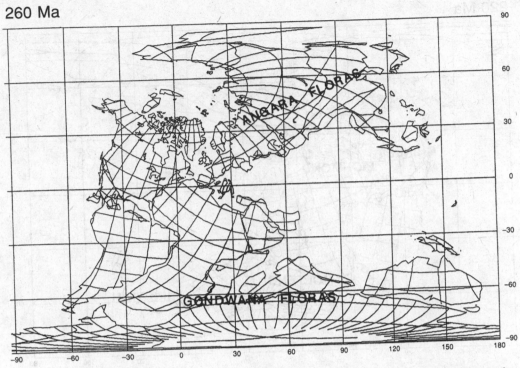

Fig. 14.1. Composite map of global continental positions for Mid-Permian Kungurian time (260 Ma), at the beginning of the period of serious angiosperm origin search, and after the disappearance of the Euramerican tropical floras of Carboniferous and early Permian times; Gondwana floras in 40–70° Permian S, and Angara floras in 30–70° N (map courtesy of A. G. Smith, 1992).

with their prey, but just recovered to dominate again the new and different Mesozoic seas. Life on the continents appears to have evolved more steadily, although with an apparently incomplete food-chain pattern (see Chapter 7); late Permian floras of Asia, particularly of China and Siberia, are likely to produce significant new records.

14.6   Triassic diversity

On land in Triassic time, and particularly towards the end of that period, there appears to have been more seed-plant diversity than for nearly 100 million years subsequently. This diversity is manifest in numerous floras found in middle paleolatitudes in both hemispheres (Fig. 14.2), includes some strikingly new characters of plants, and has led to several claims of Late Triassic angiospermy; *Sanmiguelia, Sturiella, Classopollis* and the Crinopolles group of Cornet (1989*a*) are all cases of occurrence of a single organ or of a new and unclassifiable plant showing one or more 'advanced' or unexpected characters. Late

220 Ma

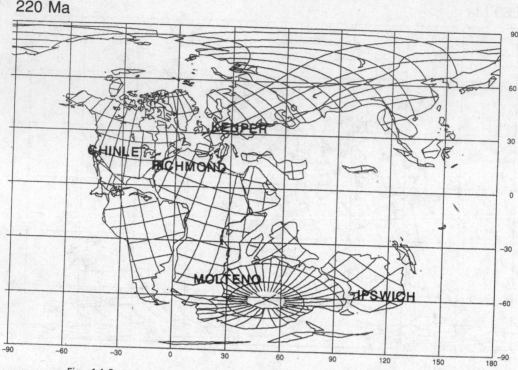

Fig. 14.2. Paleocontinental map for Triassic Early Norian time, 220 Ma, when Late Triassic floras of great diversity flourished in North America and Europe in 10–35° Triassic N, and perhaps separately in southern continents in 30–60° S (map courtesy of A. G. Smith, 1992).

Triassic time also saw the first plants with flowers, including the Bennettitales of the Williamsonia group. The function of these innovations must have been effective pollination, and as the plant organs concerned were relatively large the attracted animals were presumably small reptiles on the ground, rather than insects in flight. The new structures were not in any way connected with actual carpel and stigma angiospermy, the appearance of which was still probably 100 million years ahead in Cretaceous time.

14.7   The 'Mesophyticum'

Many Russian and German paleobotanists have used the term 'Mesophyticum' for a land-based era of slightly shorter duration than the Mesozoic. Although superfluous in stratigraphy, this extra term is descriptive of a long distinctive phase of land plant consolidation without any very prominent or sudden change, from Triassic through to mid-Cretacous time. It so happens that the mid-latitude Yorkshire Bajocian flora (Fig. 14.3) characterises the middle of this time, and is one of the finest, best-preserved and most fully worked fossil floras anywhere

Period Classification 160

	Est. no. of species					
	Leaf	♂	P	♀	S M	Stem
PTERIDOPHYTA						
Equisetites and others	8					
Lycopodites group	1		✓			10
Todites group	6		✓			1
Phlebopteiris and others	8		✓			
Klukia/Stachypteris	2		✓			
Coniopteris group	11					
Other pinnates	4					
POLLENIFERA						
Nilssonia group	29	4		2		
Williamsonia group	25	5	✓	4		4
Cycadolepis/Bennetticarpus	8			3		
Williamsoniella	3	2		2		
Caytonia	2	3	✓	3	1	
Pachypteris	2	1				
Czekanowskia group	11	1		1		
Fanleaves	14	1				
Linearphylls	6		✓			
Brachyphylls	9		✓			
Total	149					

Fig. 14.3. Period Classification 160 of Jurassic Callovian time embracing species of Pteridophyta and Pollenifera of Bajocian–Bathonian age, principally from the Yorkshire Deltaic Flora. (Information transferred from Harris (1961–1979), and other sources in the reference list.)

in the world. Its quality of preservation conceals the fact that several of the important groups of fossils from Yorkshire that may also occur elsewhere derive *all* their detailed information from this one flora. This has led to a failure to appreciate that other less well-preserved occurrences (e.g. of *Caytonia*) in space and particularly in time may not display the generally advertised features and may in fact be different in important respects, although still recorded under the same name.

The greatest reptile exuberance in size of individuals and a very extensive diversity of insects characterise the lands of about the time of the Jurassic/Cretaceous boundary. Plant life was also rich but appears to have been strongly innovative only among Pollenifera in *Cycadeoidea*, and in *Karkenia*, although this was in contrast to the stronger and less publicised radiation of pteridophytes (see Chapter 11). There was no special development of any feature subsequently to be regarded as angiospermid.

14.8   The use of Period Classifications

As first mentioned in Chapter 4, the classification of fossils and indeed also of plant reconstructions raised from them, is constrained to embrace only those fossil plant organs recorded from the interval between the selected periods and bounded by the number or name employed (see Section 5.8). Continuing the use of a taxonomic group name from an earlier period is acceptable, but no name from any later period can be extended back; thus ultimately, but in no way immediately because of the time necessary to elaborate them, classifications will come to represent successional evolutionary development with a greater chance of acceptance and stability. As presented provisionally here the Period Classifications are confined to Pteridophyta and Pollenifera and are by no means yet complete; they are designed to be added to, amplified in detail and represented whenever significant extra material has accrued.

The Period Clasification 160 (Fig. 14.3) is an example of a framework based mainly on megafossils and the dispersed palynomorphs of north-west Europe for the numbers of species but more broadly for the list of groups in which no later or particularly Holocene names are employed. The subsequent Period Classification 140 (Fig. 14.4) covering the latest Jurassic and Berriasian ages is of necessity based on more scattered information but includes the earliest (Berriasian) Wealden fossils of Watson and Sincock (1992). The differences in content and in balance from the previous classification are believed to be real but should be read in the context of first experimentation with presentations of this kind; there is virtually unlimited scope for improvement through assessment of much more widespread phenomena on a global scale. The (next) Period Classification 120 (Fig. 14.5) has so far an even more scattered megafossil basis without a major documented flora, but perhaps correspondingly more palynologic attention. The incomplete nature of this presentation is obvious but may thereby stimulate revisions and additions to an enigmatic interval overshadowed by the arrival and radiation of MCT and *Ephedripites* pollen.

14.8 The use of Period Classifications

Period Classification 140

	Leaf	♂	P	♀	S	M	Stem
			Est. no. of species				

P T E R I D O P H Y T A

	Leaf	♂	P	♀	S	M	Stem
Equisetites	3	✓					
Lycopsids		✓					
Todites group		✓					
Matonidium and others	3	✓					1
Ruffordia	2	✓					
Onychiopsis and others	3						
Other pinnates	7						
Tempskya							1

P O L L E N I F E R A

	Leaf	♂	P	♀	S	M	Stem
Nilssonia group	1	1			1		
Williamsonia group	16	1	✓	7	1		2
Cycadolepis	5						
Nilssoniopteris	1						
Cycadeoidea							1
Sagenopteris	1		✓				
Phoenicopsis/Leptostrobus	1	1		1			
Fanleaves	1						
Linearphylls	4		✓				
Brachyphylls	7		✓				
'Classopollis' plant			✓				
Witharnia	1						1
Pentoxylon	1	2		1			1
'Eucommiidites' plant			✓	1			
Total	57						

Fig. 14.4. Period Classification 140 of Cretaceous Berriasian time, embracing Pteridophyta and Pollenifera of latest Jurassic and earliest Cretaceous floras.

Period Classification 120

	Leaf	♂	P	♀	S	M	Stem
			Est. no. of species				
P T E R I D O P H Y T A							
Equisetites	1		√				
Lycopodites	1		√		√		
Matonidium	1		√				
Onychiopois	1						
Other pinnates	1						
'Water ferns'					√ 3		
Tempskya							1
P O L L E N I F E R A							
Nilssonia Group	1						
Williamsonia Group	5		√				2
Cycadeoidea Group		1		1			2
Ktalenia	1			1			
Karkenia Group	1	1		1			
Linearphylls			√				
Frenelopsis	1	1	√				
Brachyphylls							
Pseudofrenelopsis	1	1	√				
'Eucommiidites' plant			√				
'Ephedripites' plant			√				
MCT retisulc			√				
MCT crochet			√				
MCT perfotect			√				
MCT croton			√				
Total	15						

Fig. 14.5. Outline Period Classification 120 of Cretaceous Early Aptian time embracing Pteridophyta and Pollenifera megafossils, and palynologic evidence from principally Barremian time.

14.9   Cretaceous floral changes

The varied group of Barremian–Aptian monosulcate (MCT) pollen (see Chapter 9) includes from reticulate semi-tectates to almost completely tectate examples, some with columellae long or short or vestigial, some bimodal lumina and occasional trichotomosulcates and zonosulcates, some with bordered aperture. The almost complete lack of related or even of relatable megafossils may simply mean that the plants concerned had no other advanced features and are not therefore easily detectable from the existing background. The term MCT is a convenience and should not be taken to suggest a cohesive group of related plants. It is simplest to regard these monosulcate tectates as transitional to tectate triaperturates of Aptian–Albian time, although the morphologic changes required are not likely to be observed directly. Functionally, if triaperturate pollen indicates a fully developed stigmatic surface, elaboration of monosulcate tectates must indicate only rather general extra-ovular pollen germination, which was soon to be superseded and which would not attract any morphologic attention as a transitional state. At the Berriasian Period Classification 140, undeveloped monosulcate pollen (light microscope observation) presumably belonged to Fanleaves (often recorded as ginkgophytes), Bennettitales, Czekanowskiales, Nilssoniales and Cycadeoids; by Aptian Period Classification 120 MCT pollen had developed from at least some of these; by Cenomanian Period Classification 90 and certainly by Period Classification 80 comparison, most MCT types had disappeared. Subsequently around 60 Ma other different types of monosulcate appear, but it is important to record that a single sulcus is the standard simplest state and thus its presence has little positive significance.

In mid-Cretaceous times (Fig. 14.6), although the approach towards the anti-glacial maximum continued, possibly with some minor temperature fluctuations on the way, there was a final emergence of successful angiospermy in the full sense in a steady and deliberate manner through Albian to Turonian time. This was marked by universal and rapid elaboration of triaperturate pollen, indicating the new presence of a fully functional stigma or germination pad. Perhaps surprisingly no corresponding contemporaneous general radiation of physical process or of reptile or insect presence has been demonstrated. Thereafter steady radiation of angiosperms followed through to late Cretaceous time.

14.10   Cenozoic speciation explosion

The End-Cretaceous and Early Cenozoic explosion of speciation of land plants and of insects appears to have been in response to gradual re-establishment of a sharper latitudinal temperature gradient in a sequence leading through to glaciation in Quaternary time and resulting in availability of more and more diverse habitats. Land reptiles appear to have suffered a thermal control crisis in Late Cretaceous continuing into anti-glacial times (enlivened by bolides, according to taste), and cephalopods in the sea a (thermal) buoyancy crisis at about the same time; this latter was also signalled clearly in evolution of some planktonic foraminifera.

120 Ma

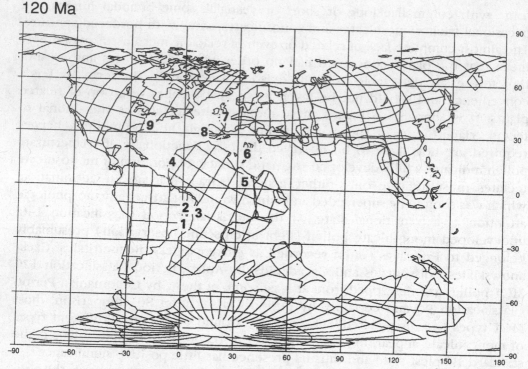

Fig. 14.6. Paleocontinental map for Cretaceous Aptian time (120 Ma), indicating published reference to occurrences of MCT pollen of Hauterivian–Aptian age. General locations: 1, Brazil, Espirito Santo; 2, Brazil, Sergipe; 3, Gabon; 4, Senegal; 5, Egypt; 6, Italy; 7, England; 8, Portugal; 9, Maryland Potomac (map courtesy of A. G. Smith, 1992.)

Following the inevitable extinctions of K/T boundary fame, mammals and fish respectively soon replaced the reptiles and cephalopods, and plankton was reorganised. Land plants benefited directly from all the physical changes mentioned and thus do not display extinctions at the K/T boundary any more than they did at the Permian/Triassic boundary. Indeed their development moved on towards renewed establishment of most varied tropical and other habitats of Holocene time (Doyle 1978a), for the first time since Carboniferous and Permian times recognisable tropics were recorded in the Euramerican Coal Measures.

14.11   Major geologic events

Although popular as dramatic concepts supposedly affecting certain aspects of biological evolution, neither widespread intense vulcanism nor impacts of extra terrestrial bodies have yet been shown in any particular case to have had any truly global and enduring effect. Changes of sea-level have to be global but were probably entirely continuous and slow enough to be compensated for locally by sedimentation. Tectonic plate destruction in critical areas, although irreversible, was almost certainly too slow to influence organisms in general. Magnetic reversals, although not proven to have had any biological effect, appear to have been of unexplained pattern and of inadequate intensity. This leaves the only serious probable cause of either significant or catastrophic biological change as one-way climatic change in paleolatitudes where coasts and other major topographic features happened to be arranged in an east–west direction. It is thus difficult to suggest convincing geologic control of a phenomenon of such ultimate importance as angiosperm innovation; perhaps like much in human history it was subtle enough to have passed unnoticed at the time, so that precise detection is beyond the degree of resolution available.

15 General biological conclusions

15.1 Grade of development

A conclusion of this review of the angiosperm origins problem is that angiospermy is now best considered independently of all current taxonomy as a grade of development achieved by several of the principal Pollenifera (seed-plant) groups by mid-Cretaceous time (Fig. 15.1). Consequently an agreed convention is required as discussed in the next two sections. This is clearly contrary to a tradition of definition that has persisted for many decades with little change, presumably because it was locked on to the concept of an evolutionary event of probable monophyletic nature.

15.2 'Older' characters

It has been customary to refer to a list of important characters by which angiosperms may be distinguished from all other plants (see e.g. Heywood 1978). The list, however, cannot be considered as satisfactory in that some characters are carried by other plants as well, and numerous accepted angiosperms lack one or more of the characters. It is therefore profitable to divide the list into 'older' characters present in one or more plants of Jurassic and earlier times, and the less common 'newer' characters presumed to have appeared only in Cretaceous time.

Late Paleozoic Carboniferous and Permian times saw the development of both the megaspore retained to become a seed in several groups including the Lepidodendraceae, and the monosulcate pollen in the Medulloseae and other groups. As early as Late Triassic time flower structures were characteristic of early members of the Williamsonia group and of *Sanmiguelia*; monosulcate tectate pollen of the group Crinopolles was described by Cornet (1989*a*), after a much earlier record by Schulz (1967), although no suggestion has been made of a function of such featues. The leaf lamina and petiole were well developed in both pteridosperms and ferns; venation was variously anastomosing in *Glossopteris*, *Lonchopteris* and others, and even reticulate in ferns and in several species of *Furcula* from different areas. In Jurassic time the closure of several ovules within a carpel-like structure characterised *Caytonia* and *Lepidostrobus*, and it seems likely that pollen tube germination took place outside the micropyle and pollen chamber. Thus, all these 'older characters' had already been

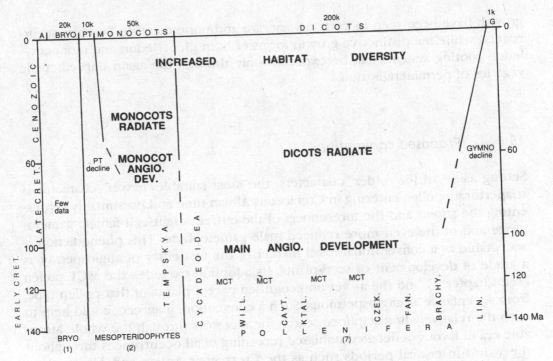

Fig. 15.1. The grade of development view of angiospermous plants in mid-Cretaceous time. Vertical scale in millions of years before present (Ma). Horizontal scale at Holocene in approximate numbers of species; (k = 1000) horizontal scale in Cretaceous representing proportions of them existing land flora. Cretaceous abbreviations as used previously. A, terrestrial algae; ANGIO., angiosperm; BRYO., Bryophytes; PT, Pteridophytes; MCT, monosulcate columellate–tectate pollen; WILL, Williamsonia group; CAYT., Caytonia/Sagenopteris Group; KTAL., *Ktalenia*, CZEK., Czekanowskia group; FAN., Fanleaf group; BRACHY., Brachyphylls; LIN., linearphylls; G, gymnosperms.

incorporated individually into pre-Cretaceous plants that were not otherwise obvious claimants to be considered to be angiospermous.

15.3 'Newer' characters

In contrast there are observable in fossils two major character sets that have no history of record before mid-Cretaceous time. First, the development of a stigma (Thomas 1934) and the corresponding triaperturate or polyaperturate pollen, presumably associated with the greatly reduced gametophytes, form a set observable mainly through the fossil pollen, but also sometimes in mesofossil flowers. Secondly, vessels first appear in Cenomanian wood, and although this date may be varied slightly by new records, it is unlikely that significantly earlier fossils have been overlooked. Two further character sets are far less certain to be revealed in fossil evidence: the embryo, cotyledons, endosperm, and early growth have been seen in certain very rare and unpredictable Tertiary occurrences, while the distinctive growth forms of both dicotyledon and monocotyledon rooting systems can be expected only through the again unpredictable vagaries of permineralisation.

15.4   Proposed convention

Setting aside all the 'older' characters, the most suitable 'newer' character is triaperturate pollen entering in Cretaceous Albian time and presumably representing the stigma and the forerunners of the current eight-cell female gametophyte and of the even more reduced male gametophyte. This phenomenon is acceptable as a conventional fossil marker of the first entry of angiospermy as a grade of development of seed-plants. Its adoption excludes the MCT pollen (see Chapter 9), and the as yet unrecognised parent plants of that pollen type, from acceptance as angiospermous. Such a convention, if agreed, could help to curb the relatively undisciplined search for ancestors through the whole Mesozoic era in favour of detailed unbiased recording of all occurrences throughout the world in critical periods such as the Cretaceous Aptian and Albian. Even when focussed recording has been attempted, much greater effort has then been applied to the interpretation of the American and African equivalents of the English MCT pollen (see e.g. Doyle and Donoghue 1987, 1992, 1993, Donoghue and Doyle 1989, Doyle et al. 1990a,b, Doyle and Hotton 1991), and discussion of extant family associations has thereby been driven along. These papers, however, exhibit a plethora of microhypotheses and alternatives, breathlessly based on shuffling of characters to and fro between the authors and their parsimonious 'trees'. Characters (with few available to record) of selected fossils are analysed together with the full characters of whole extant groups in a shared search for apomorphies. To keep the matrices to economic workable size, single character states have to do service for whole groups, with all monocotyledons placed in one file as an extreme example.

In selecting 'trees' for parsimony no account is taken of geologic time or the order of occurrence of the taxa concerned. Anxiety or occasionally dogma is expressed over the monophyletic, paraphyletic or polyphyletic nature of groups. Fossils are still regarded as merely useful for testing (Doyle and Hotton 1991, p. 171) the primary hypotheses essentially derived from comparative morphology, with monophyly as a complete methodological constraint.

No doubt the statistical aspects of the manipulation procedures are well respected, but the question to be answered concerns the successful outlining of an evolutionary sequence of occurrences through the critical stages of the Mid-Cretaceous time-scale. It seems preferable to regard this matter as still very open and thus waiting for the time some way ahead when details of fossil occurrences accumulated on a scale scarcely yet envisaged will have made sense of the sequences. There is no victory to claim right now.

15.5   *Mesozoic Pollenifera and dicotyledonous angiosperms*

This Mesozoic seed-plant group has been presumed for many years to have been ancestral to various groups of angiosperms, one or more according to taste; unless the angiosperm ancestors were cryptic and therefore totally unknown, which is unlikely, one or more members of the Mesozoic Pollenifera must have been involved. It is perhaps worth recalling again that the pollen of all serious Mesozoic candidates was monosulcate. Because so many dicotyledon angiosperms have triaperturate pollen, the origin of this pollen must have been achieved very rapidly in Aptian–Albian time and must have been indicative of fossil closure of some carpels at that time. The actual transition from monosulcate to simple triaperturate was perhaps quite easily achieved and not therefore suitable for lengthy theorising about a mechanism, or search for any single transition. The most likely plant megafossil candidates for ancestry do not offer any obvious transitions towards the newly discovered Cenomanian mesofossils (Fig. 15.1).

The *Caytonia* group, despite having its sole information base back in Jurassic Bajocian time (before 160 Ma), has early Cretaceous representatives and suitably disappears from the fossil record in Mid-Cretaceous time. Very little is known of its post-Bajocian reproductive structures; its very small pollen is monosulcate, and its saccate features would have disappeared quickly when no longer functional for micropyle flotation. Oddly, no thus-far discovered Cenomanian angiospermous plant suggests any close connection with *Caytonia* or *Ktalenia*; this is perhaps because the 60 million year gap in the knowledge of reproductive structures precludes considering it as any more than a likely candidate.

The *Leptostrobus–Ixostrobus–Phoenicopsis* plant is known best from Early Cretaceous records, but also appears to have persisted after Cenomanian time in north-east Asia. At present it appears to be the most likely candidate to have given rise perhaps to the 'Cenoham' 92 group including *Spanomera* and *Caloda* among current 'Cenomanian' discoveries with a supposed hamamelid affinity, and thus to some subsequent Late Cretaceous radiation.

Williamsonia is the only long-ranging member of the Mesozoic Pollenifera to produce the actinomorphic flowers that continued to exist in some considerable variety in Early Cretaceous time (Watson and Sincock 1992); *Cycadeoidea*, which may not be at all closely related to *Williamsonia*, is virtually an Early Cretaceous phenomenon; and *Williamsoniella* is too isolated and remote in time to consider. Apparently angiospermous actinomorphic flowers made a slow appearance, e.g. *Asterocelastrus* and *Kalinaia*, in mid-Cretaceous time, and were not large. Such symmetric flowers were the most obvious type of attractant for flying insects and could well have been evolved several times.

As so often happens in geology and paleontology, the numerous well-documented Cenomanian discoveries of the last ten years are so informative in themselves that the linkage problem, even if the idea of monophyly is still adhered to, has become more, and not less, difficult to understand.

Although the genus *Gnetum* of living tropical plants (mostly lianes) is frequently mentioned in connection with angiosperm origins, there is as yet no fossil record and no realistic prospect of discovering fossils other than dispersed pollen. The presence of leaves with reticulate venation, and some vessels in the wood, make *Gnetum* no more and no less remarkable than several isolated taxa counted as angiosperms, although often with an 'incomplete' set of characters. The pollen of *Gnetum* is very small and spherical, monoporate, with no obvious parallels in other taxa and with few distinctive features of its own. There is no logical reason to give the members of this small group any priority of thought merely because they appear to be merely too primitive to count as angiosperms and are consequently regarded automatically as gymnosperms.

15.6   Origin of Holocene gymnosperms

This very small although locally significant group of plants in higher latitudes may best be considered as a survival of only a minor part of the Mesozoic Pollenifera. Some of its members, as seen in geologic time, appear now to be close to extinction, e.g. many Cycadales, Taxodiaceae, Araucariaceae, *Ginkgo*, and *Welwitschia*. Even the few larger groups of Pinaceae, Cupressaceae and Podocarpaceae reveal together less diversity than a single major angiosperm family. It is now perhaps better to regard the few living gymnosperms as marginal evolutionary products of the Mesozoic Pollenifera; hitherto the automatic inclusion of all living gymnosperms and Mesozoic Pollenifera under one head as Gymnospermophyta has seriously distorted interpretation of the fossil record and indeed of the origins of angiosperms.

As an example, the genus *Ephedra* contains living plants centred on warm temperate semi-arid regions, with some late Neogene attributed megafossils; its fossilisation potential is low. The polyplicate pollen is distinctive and is only paralleled among living plants in monospecific *Welwitschia*, which may possibly be related and for which the fossilisation potential is extremely low. It seems likely that the various different polyplicate pollens of Late Triassic and of Mid-Cretaceous age in particular represent other plants that may not be related; Mid-Cretaceous *Drewria* may be one of these. It is not clear what kind of ordinary plant might be sought as an ancestor of the several very specialised living *Ephedra* species. Although most other extant gymnosperms are more confidently associated with selected ancestors among Mesozoic Pollenifera, rather few cases could be considered as adequately confirmed by fossil succession because it has never appeared necessary or interesting enough for the assumptions involved to be questioned.

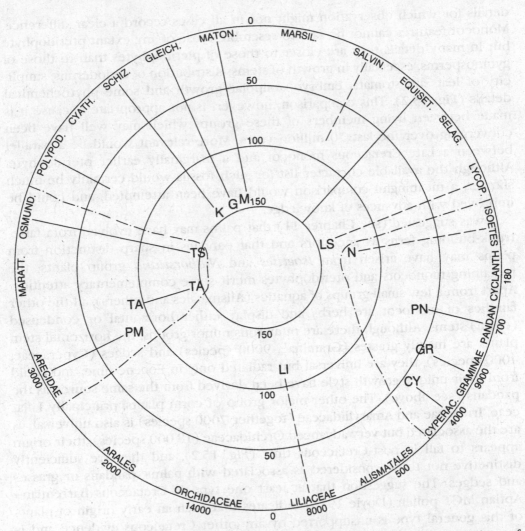

Fig. 15.2. Some possible relationships in Cretaceous and Tertiary time of Pteridophyta and monocotyledons. Circles are labelled in millions of years before present. Radial lines indicate some continuity of presence of a group. Fossil occurrences: CY, Cyperaceae; G, *Gleichenia* group; GR, Graminae (grasses); K, *Klukia*; LI, Liliaceous pollen (MCT); LS, *Lycopodites*; M, *Matonia* group; N, *Nathorstiana*; PM, fossil palms; PN, fossil pandans; TA, *Tempskya*; TS, *Todites*; MARATT., Marattiaceae; OSMUND., Osmundaceae; POLYPOD., Polypodiaceae; CYATH., Cyatheaceae; SCHIZ., Schizeaceae; GLEICH., Gleicheniaceae; MATON., Matoniaceae; MARSIL., Marsiliaceae; SALVIN., Salviniaceae; EQUISET., Equisetaceae; SELAG., Selaginellaceae; LYCOP., Lycopodiaceae; CYCLANTH., Cyclanthaceae; CYPERAC., Cyperaceae; PANDAN., Pandanaceae.

15.7 Origin of monocotyledonous angiosperms

It has usually been possible to draw strong clear contrasts between all dicots and all monocots, leaving only flowers, closed carpels and double fertilisation details for which observation might not in all cases record a clear difference. Monocot features cannot be said to resemble those of any extant pteridophyte but in many details they are closer to those of pteridophytes than to those of gymnosperms, especially in growth of stems, distribution of endodermis, simplicity of leaf and stomata, embryo, unipolar growth and some phytochemical details (Fig. 15.2). This comparison, however, is not appropriate because it is made between living members of these groups which may well have been converging over the last 70 million years. More relevant would be a parallel between a Late Cretaceous monocot and a marginally earlier pteridophyte. Although the available character list for such fossils would certainly be much shorter, a meaningful comparison would have been attempted, and could be improved with advances of knowledge.

It was suggested (see Chapter 11) that palms may have evolved from false-trunk-building ferns as ancestors and that pandans in sharp distinction from palms may have arisen from *Isoetites* and *Nathorstiana* group plants; the remaining monocots and pteridophytes merit some complementary attention. Apart from a few small groups of aquatics (Alismatales and others), all the other families of monocot are herbs and display either horizontal or condensed (corm) stems. Although there are numerous minor groups, the horizontal stem plants are mainly grasses (Graminae, 9000 species) and sedges (Cyperaceae, 4000 species); they are universal but radiated only in Eocene time, and could from their unique growth style have been derived from the same source as the pandans (see above). The other major group of corm plants, principally Liliaceae, Iridaceae and Amaryllidaceae (together 7000 species) is also universal, as are the associated but very advanced Orchidaceae (18 000 species); their origin appears to fall in post-Cretaceous time (Fig. 15.2), and they are sufficiently distinctive not to be considered as associated with palms, pandans or grasses and sedges. The suggestion that at least one type of Cretaceous Barremian–Aptian MCT pollen (Doyle 1973) indicates that such an early origin of plants of this general type is unsupported by any other Cretaceous evidence and is thus not immediately attractive. However, all pollen of 'monocotyledonous' living plants including the Liliaceae and of rather few dicots is monosulcate, or monoporate.

Finally, of the principal pteridophyte groups only the Sphenopsida have not been invoked; their sole surviving representatives *Equisetites* and living *Equisetum* appear to have occupied a permanent and unique water-margin niche leading to a long-range stable plant form, comparable with that of brackish intertidal *Lingula* among invertebrate animals; no other evolutionary developments are likely in such cases.

15.8   *Pteridophyte evolution*

A radiation of these plant groups into more diverse habitats in Late Jurassic–Early Cretaceous time has been mentioned (see Chapter 11). Further radiation

in post-Maastrichtian time into rapidy diversifying angiosperm-dominated Cenozoic habitats has produced the large group of living leptosporangiate ferns in
Holocene lower latitudes. With the exception of this large group, the remaining
pteridophytes now have a minority status comparable to that of living gymnosperms (see Section 15.6). They form a marginal evolutionary product from the
universal but not so well investigated and understood Mesozoic pteridophytes.
Hitherto the pteridophyte literature has relied overenthusiastically on incorpor-
ating the too remote although very well-preserved Late Paleozoic plant groups
with homosporous often pinnate megaphylls. In attempting to assess the suggestion of the origins of monocotyledonous angiosperms in this group, much more
detailed attention to all Mesozoic pteridophytes will be necessary; because of
the lack of thick cuticles this will be more difficult than investigating the
Pollenifera.

15.9   Animal integration

From considerations of timing alone, there is no obvious deduction of animal-led
evolution of angiospermous or of other plants. Neither Lepidoptera nor higher
Hymenoptera were prominent ahead of Late Cretaceous angiospermous plants.
Even if the more mobile Reptilia and the Birds are taken as a continuum, there
is no striking connection with plants; mammals were not seriously in contention
until Eocene time. Although all the evidence is currently weak, it may be more
profitable to allow for Cretaceous floral evolution to have been plant led (Fig.
15.3).

15.10   Future solutions

Although outstanding contributions have emanated from most world centres of
botanical study, few schools have persisted in the field of Mesozoic fossils as
long as that of A. C. Seward from before 1890, through Hamshaw Thomas to
T. M. Harris and all of their many botanical associates for virtually a century.
Even this amount of continuous discussion did not really resolve the problem
of angiosperm origins, and so perhaps there is relevance in the question of why
this should be so and whether some ingrained pattern of prejudice in the studies
referred to above and elsewhere had too great a negative influence. Any excit-
ingly simple evolutionary change would have been apparent before and would
have long since been developed in some way. The future appears to lie in
gradual mastery and integration of the whole of the available fossil record
including paleopalynology, and especially in unbiased data handling of the products. All of this is now technically possible but it will only succeed with those
who are prepared to counter the old negative influence by trying new lines of
both thought and action, however far-reaching the results may prove to be
for the currently observed so-called higher taxonomic levels of classification.

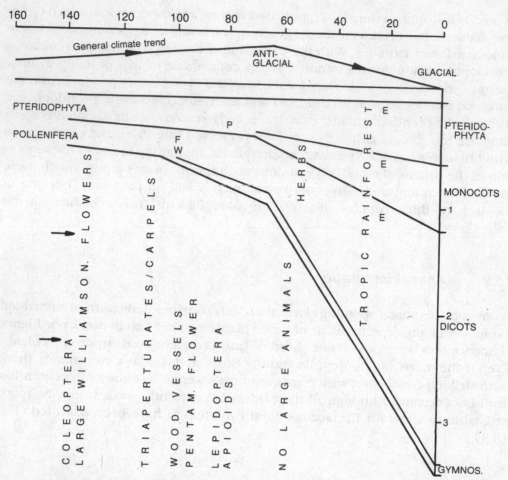

Fig. 15.3. View of the general course of events in the history of Tracheophytes from Early Cretaceous to Holocene time. The vertical scale represents approximate or guessed numbers of family sized taxa. The horizontal scale is geologic time scale in millions of years before present. Words placed vertically are to suggest influences at about the time indicated. E, epiphytes; F, flowers; P, palms; W, wood with vessels; WILLIAMSON., Williamsonia; PENTAM., pentamerous; GYMNOS., gymnosperms.

Additionally the whole subject is only likely to be considered economically worthy of funding with a different approach, possibly such as the one offered above.

Glossary

Terms are marked (Geol) if primarily employed in earth sciences or with fossils; all others are botanical or biological.

actinomorphic   Of a regular flower with all petals of the same shape.

adaptation   Name for a concept of suitability of a character of an organism for a presumed function; name considered inappropriate and thus *not* used in this work.

advanced   Of a plant character believed to have arisen late in evolution.

Advancement Index   From the proportion of advanced characters exhibited by a family (or other taxon), expressed as a percentage.

age/(stage) (Geol)   Subdivision of a geologic period, employed in describing the Global Stratigraphic Scale; 'stage' is used for the same but is incorrect for this purpose.

aggradational (Geol)   Of a land area composed of newly deposited sediment, as in a river flood-plain and delta.

alternation of generations   Gametophyte haploid (n chromosome) sexual stage is followed by sporophyte diploid ($2n$) asexual stage as an alternation, in bryophytes and all higher plants.

alveolate   Of the middle ectoexine of some pollen that has an outer tectum layer, and shows lacunae (many small spaces) but is not columellate nor even granular.

amber (Geol)   Fossil resin assumed to have been derived from pinaceous conifer trees in past geologic periods.

anastomosing   Of the veins of a leaf that fork and rejoin successively (e.g. *Glossopteris*, Fig. 8.3).

anasulcate   Of a sulcus on the distal surface of a pollen grain.

anatropous   Of ovules that are inverted through 180° in their direction of growth.

androecium   Collective term for all the pollen-producing parts of a flower.

anemophily   Process of pollination of a flower by wind.

angiosperm   Plant with flowers, with carpel enclosing the ovules, and with pollen with non-laminate endexine.

angiospermid   Of a plant character normally found only in an angiosperm.

angiospermous   Of fossils of plants that show some characters of angiosperms but of which knowledge is incomplete.

anomocytic   Of epidermal stomata lacking subsidiary cells.

anther   The head of a floral stamen with pollen sacs and connecting tissue.

anthocorm	Evolutionary concept (Meeuse 1981) of central axis becoming a group of fertile gonoclads; contrast foliar carpel theory.
anthophyte	Seed-plant with flower or comparable reproductive structure.
anticline (Geol)	Strata folded upward in the form of an arch.
anti-glacial (Geol)	Of a past era in which the earth's surface was believed to have been warmer than average, in contrast to glacial.
apocarpous	Of angiosperm gynoecia with carpels free and not fused together; cp. syncarpous.
apterygote	Of insects without wings, growing only by a series of moults; described as primitive.
arachnid	Arthropod organisms including spiders and mites.
arborescent	Of a long-lived plant with tree form.
banded (muri)	Of muri of pollen tectum showing apparently annular rings.
base-taxon	A category of taxa of fossils to include the species, the paleotaxon, and any others employed to denote units considered to be of comparable scope to the species of living organisms.
binominal	Of biological nomenclature with formal latinised generic and specific epithets in the name of a fundamental taxon.
biorecord (Geol)	Paleontologic formal fundamental taxon, now renamed a paleotaxon (see Hughes 1989).
bipinnate	Of compound leaves with pinnae of two orders.
bisaccate	Of pollen with two separate air sacs developed within the pollen wall.
bisexual	Of a flower possessing both male and female reproductive structures.
bitegmic	Of ovules with two separate integuments.
boundary point (Geol)	Point of definition in rock selected to mark the beginning of a time-scale division.
brachyphyll	Group of extinct seed-plants with short leaves closely adpressed to the stem.
bract	Leaf-like organ subtending an inflorescence.
bracteated gonoclad	Theoretical ancestral fertile structure (see Meeuse 1981).
bryophyte	Plants, including liverworts and mosses, that are primarily gametophytic and are non-vascular.
calcareous (Geol)	Rock strata of sedimentary origin composed primarily of calcium carbonate.
cappa	Proximal exinal thickening prominent in some bisaccate pollen grains.
capsule	Dry dehiscent fruit from a syncarpous ovary.
carnivore	Animal that feeds entirely on other animals.
carnosaur	Carnivorous dinosaur.
carpel	Part of angiosperm gynoecium, containing ovules; consists of ovary, style and stigma.
Cenoham 92 group	Cenomanian angiospermous fossils (as known in 1992), which have been credited as resembling hamamelids.
Cenomag 92 group	Cenomanian angiospermous fossils (as known in 1992), which have been credited as resembling magnoliids.

Cenoros 92 group   Cenomanian angiospermous fossils (as known in 1992), which have been credited as resembling rosiids.

Cephalopoda   The most advanced group of Mollusca, typically active and free-swimming, e.g. squid, octopus, ammonoids, belemnoids.

character   Feature of a plant selected and labelled for convenience in analysis or discussion.

chlamydosperms   Alternative name applied to Gnetales.

chloranthoid   Of fossil plant organs carrying some characters resembling those of one of the extant Chloranthaceae.

clastic (Geol)   Of a sediment composed of fragments of pre-existing rock.

climax   Final stable plant community established after a series of changes.

colpus   Longitudinal furrow in pollen exine, associated with germination; plural colpi.

columella   Rod-like ectexine element supporting the tectum; in angiosperm pollen.

comparison record (Geol)   Formal method of recording fossil data by comparison with a paleotaxon or with a species (see Hughes 1989).

compression (Geol)   Fossil preservation in which the organism is flattened during rock consolidation.

conformable (Geol)   Of strata that lie parallel, one above the other.

conifer   Largest group of living gymnosperms, normally with small or needle-shaped leaves.

coniferophyte   Group term for fossil conifers and other fossils resembling them in some characters; *not* used in this work.

corolla   Collective term for all the petals of a flower.

cotyledon   Leaf-like part of the developing embryo in seed plants.

crotonoid   Type of MCT (see below) pollen, and some extant angiosperm pollen, with raised supratectal elements in a distinctive pattern of triangles and rectangles.

crustal (Geol)   Of the crust of the earth, as distinct from the upper mantle and other layers within.

cupule   Structure enclosing at least partly one or more seeds, in the pteridosperm group of fossil plants.

cuticle   Waxy layer that impregnates the outer wall of epidermal cells, acting as a water-seal.

cuttings (Geol)   Fragmentary rock material produced in making a borehole; used for study when no core has been retrieved.

cycad   Pantropical group of ten living gymnosperm genera with large cones and large pinnate leathery leaves.

cycadeoid   Group of Mesozoic (mainly Cretaceous) plant fossils bearing flowers, typically large and permineralised.

cycadophyte   Group term for fossil cycads and other fossils resembling them in some characters; *not* used in this work.

deciduous   Natural shedding of leaves, frequently seasonal.

dehiscence   Natural splitting open of a fruit to release seed on ripening.

dentition   Here for the arrangement of teeth in the jaws of a vertebrate animal.

diagenetic (Geol) Of the physical and chemical processes undergone by sediments during their consolidation under gravity to form rock.

diaspore Collective term for spores and seeds dispersed from a plant.

dichotomous Of leaf venation that shows symmetric equal forking.

dicot Abbreviation commonly used for dicotyledon.

dicotyledon Plant defined by possession of two seed-leaves (cotyledons) on germination of the seed.

dinocyst (Geol) Fossils believed to represent the preservable cyst stage of dinoflagellate algae.

dinoflagellate Group of motile aquatic unicellular algae in the size range of 10–200 μm.

dinosaur Members of several extinct Mesozoic orders of reptiles.

dip (Geol) Inclination from the horizontal of a rock stratum, as measured in the field.

distal pole Of a spore or pollen grain, on the outside away from the proximal pole that is adjacent to the tetrad centre of origin.

double fertilisation Process characteristic of angiosperms, one male nucleus to diploid zygote and another to triploid endosperm.

ectexine Outer of two layers of pollen wall exine.

elaterate (Geol) Of palynomorphs bearing attached elaters found in paleotropical Cretaceous Albian rocks.

elytra Hardened wing cases of Coleoptera (beetles); singular elytron.

emendation Formal alterations to the name and/or circumscription of a taxon made after its original description.

endemic Of organisms only found in a designated area or region.

endexine Inner of two layers of pollen wall exine, usually homogeneous.

endodermis Distinct single layer of cells at inner boundary of cortex against the stele; visible in all roots and in pteridophyte stems.

endopterygote Of winged insects with complete (holometabolous) metamorphosis including a pupal stage; larvae entirely different from adults; regarded as the most advanced insects.

endosperm Tissue for nutrition surrounding the embryo in angiosperm seeds ($3n$ chromosomes).

entomology Study of insects.

entomophilous Of plants pollinated by insects.

epiphyte Plant growing on limbs of a larger plant without parasitism, e.g. many tropical orchids and ferns.

era (Geol) Large (long) division of geologic time-scale, e.g. Mesozoic era.

eustatic (Geol) Of global changes of sea-level caused primarily by formation or by melting of ice-caps on land.

evaporite (Geol) Chemical deposits formed by precipitation on evaporation of natural bodies of water.

Evolutionary Distance (ED) Concept (Chapman 1987) in development of Advancement Index, measuring separation of families from their character differences.

exine	Outer sporopollenin part of pollen wall; distinguished from intine, which is without sporopollenin.
exopterygote	Of winged insects with simple incomplete metamorphosis (no pupal stage); immature stages known as nymphs, or larvae.
extant	Of all organisms living at the present day.
extinction (Geol)	Moment of time when all individuals of a species have died without issue.
facies (Geol)	Sum of characters of lithology and fossil content used to express the nature of a rock formation at one locality for comparison with others.
false-trunk	Tree trunk produced from accretion of many rootlets round a relatively slender true stem, e.g. in tree-ferns.
fault (Geol)	Large-scale fracture of crustal rocks, with detectable displacement.
filament	Immediate stalk of an anther in a stamen.
florule	A locally developed flora, considered to be a facet of a larger regional flora.
flower	Assembly of reproductive structures on a plant, usually with an attractant (visual or olfactory).
follicle	Fruit consisting of carpel with seeds, dehiscent along one side.
fructification	Collective term for aggregated fruits of a seed plant.
fruit	Ovary with seeds in mature state.
fusainised (Geol)	Of fossilised plant material reduced to carbon, often as a result of natural fire before burial.
gametophyte	Haploid gamete-producing phase of life-cycle; see alternation of generations.
gamoheterotropy	Transfer of plant characters in evolution from one sex to the other.
gekko	Family of small lizards with adhesive digits.
gene-pool	Sum and variety of all genes present in a breeding population at one time.
genus	Name for the first aggregation of species into the lowest level of classificatory hierarchy, below 'family'.
geochronometry (Geol)	Technique of making radiometric time determinations.
geologic period (Geol)	Large division of global geologic time-scale of duration 10–100 Ma, e.g. Jurassic period.
glacial (Geol)	Of a past or present era in which the earth's surface is known or was believed to be colder than average, inducing formation of land ice.
glauconite (Geol)	Green iron silicate mineral of sedimentary rocks, containing potassium used in radiometric dating.
glycerine jelly	Material originally used for mounting palynomorphs on glass slides for light microscopy; superseded because of disadvantages.
gnetalean	Group term for all species of extant *Gnetum, Ephedra* and *Welwitschia*; presumed to be closely related.
gnetophyte	Group term for fossils bearing some characters of living Gnetales; *not* used in this work.
gonoclad	Conceptual male or female ancestral organ (Meeuse 1989).

gonophyll	Conceptual ancestral organ, contrasted with carpel (Melville 1983).
graded comparison (Geol)	Formal style of comparison record, in place of unqualified attribution; see Hughes (1989).
granular	Of the middle part of a pollen ectexine below the tectum; material in section appears to be in distinct granules.
gymnosperm	Extant seed-plant with ovules and thus seeds 'naked' (not enclosed in an angiosperm carpel), e.g. living conifers and cycads.
gynoecium	Term for the whole pistillate or female part of a flower, including carpels and ovules.
hadrosaurs	Duck-billed dinosaurs, of the Ornithischia.
Hamamelidae	Subclass of dicotyledonous angiosperms with simple flowers, including the trees plane, birch and walnut.
haplocheile	Of stomata, in which guard cells and subsidiary cells develop from different epidermal cells.
harmomegathic	Of adjustments of the wall to volume change of pollen during dehydration and rehydration.
haustorial	Of organs capable of sucking fluids.
herbivore	Animal feeding entirely on plant material.
Holocene	Latest period of the Stratigraphic Time-Scale, commencing 10 000 years ago, and including the present day.
holotype	Single fossil specimen selected as a permanent nomenclatural reference, from material used in the original description of the base taxon concerned.
homologous	Of characters having the same phylogenetic origin.
ICBN	International Code of Botanical Nomenclature; latest version by Greuter *et al.* (1988).
imago	Adult (reproductive) phase of an insect life-cycle.
impression (Geol)	Preservation of fossil with all organic material removed from a compression, leaving only a print on the rock surface.
infauna (Geol)	Fauna of animals living buried within sediment.
inflorescence	Aggregation of numerous flowers.
infructescence	Aggregation of numerous fruits, derived from an inflorescence.
insectivore	Animal feeding principally or wholly on insects.
integument	Vascularised envelope enclosing an ovule.
interseminal scale	Sterile appendages set between the ovules in plants of Bennettitales (fossil).
isostatic (Geol)	Of vertical movements of areas of the earth's crust.
K-selected	Of the reproductive 'strategy' of species characteristic of stable environments.
K/T boundary	Cretaceous/Tertiary boundary.
laesura	Trace or scar on proximal surface of spore, indicating original contact with other members of a tetrad.
larva	Early non-reproductive phase of an insect or other invertebrate life-cycle.
Laurasia (Geol)	Group term for reference to phenomena applying to North America, Europe and Asia; employed for Mesozoic records.

lens (Geol)	Term for a stratum that is observed to thin in all directions laterally and therefore to be absent from other rock successions.
leptoma	Thin region of pollen grain at distal pole, associated with germination (emergence of pollen tube).
lignite (Geol)	Product of diagenesis of peat, a low grade or brown coal.
lignitised (Geol)	Of single compressed plant fragments at the lignite level of diagenesis.
lineage	Ancestry, usually expressed in terms of base-taxa believed to be successors in time.
linearphyll	Collective term for all Mesozoic Pollenifera bearing small or linear leaves reminiscent of those of certain living conifers.
lithology (Geol)	Includes all descriptive features of a crustal rock; sometimes restricted to petrologic information.
littoral (Geol)	Of sedimentary deposits, taken to represent a sea-margin.
locality (Geol)	Recorded present-day location of rocks under study.
lumen	Space between muri (walls) of a reticulum when present in semi-tectate pollen; plural lumina.
'lumping'	Practice of combining taxa for easier interpretation purposes; should not be permitted to affect the form of the database.
Ma (Geol)	Abbreviation for a million years before the present day, used with a number.
Magnoliidae	Subclass of dicotyledonous families, believed to be primitive; some fossils have been attributed to this group on probably insufficient evidence thus far.
mangrove	Plant habitat on present tropical tidal marine margins of aggradational land.
marker-point (Geol)	Selected and designated point in a rock succession, to mark the beginning of a Stratigraphic Time-Scale division.
MCT	Monosulcate columellate–tectate pollen of unknown affinity from Cretaceous Hauterivian to Aptian rocks.
megafossil (Geol)	Fossil that is at least provisionally identifiable using a hand-lens.
megaspore	Larger spore of a heterosporous plant; in seed-plants, retained in the ovule.
megasporophyll	Leaf-like organ bearing megaspores or ovules.
meiosis	Reduction division of a diploid ($2n$) cell to form four haploid (n) cells, leading to production of gametophyte phase. See 'alternation of generations'.
meristem	Sector of plant tissue of unspecialised cells from which specialised tissues are systematically formed.
mesarch	Of developing protoxylem wood first formed during elongation of the plant organ, which is seen to be surrounded by later-formed metaxylem indicating growth both centripetally and centrifugally in the stem.
mesofossil (Geol)	Relatively new term for intermediate-size plant fossils such as megaspores and small seeds, visible in the field but requiring microscope study.
mesophytes	Of plants living in temperate climates with average rainfall.

Mesophyticum (Geol)	Name of an era based on plant fossil information, cp. Mesozoic; only employed seriously in the former USSR and Germany.
microflora (Geol)	Assemblage of dispersed pollen and spores buried in a sedimentary rock, not necessarily of plants living together.
microfossil (Geol)	Fossil requiring microscope study, usually not visible to the naked eye in rocks in the field.
micropylar drop	Exuded fluid drop that may trap pollen, to be sucked through the micropyle to the pollen chamber.
micropyle	Opening through the integument at the apex of an ovule, leading from outside to the pollen chamber to allow pollination.
microspore (Geol)	Smaller spore of a heterosporous plant; in seed plants represented by the male gametophyte in pollen.
microsporophyll	Leaf-like organ bearing microspores or pollen.
miospore	Collective term for small fossil dispersed spores and pollen of undetermined affinity.
monocolpate	Pollen grains with single distal sulcus; monosulcate is preferred because of definition of colpus (see above).
monocot	Abbreviation commonly used for monocotyledon.
monocotyledon	Angiosperm; usually with parallel veined leaves, floral parts in threes and one cotyledon with the embryo.
monoecious	Of plants bearing both pollen- and ovule-producing organs on the same individual.
monolete	Of spores with single straight laesura, suggesting origin in a vertical or parallel tetrad; as distinct from tetrahedral tetrads with trilete spores.
monophyletic	Of a group of related plants assumed to have one ancestor.
monoporate	Of pollen grains with only one distal pore.
monosaccate	Of pollen with a single saccus extending all round the grain at the equator or just distally.
monosulcate	Of pollen grains with one distal sulcus or furrow.
mummified (Geol)	Of a compression plant fossil, lightly consolidated and excellently preserved; examples in Jurassic rocks of East Yorkshire.
muri	The walls of a reticulum if present in a semi-tectate pollen grain; singular murus.
nannofossils (Geol)	Fossils of organisms believed to have been nannoplankton.
nannoplankton	Very small planktonic plants such as coccolithophores.
natural selection	Selection for eventual breeding, by the effect of outside agencies, of certain juvenile organisms from an original large production.
natural survival	Survival by chance to breed of a few individuals of an original extensive reproductive process; preferred here to 'natural selection'.
nectary	Nectar-secreting gland, usually in a flower; for attracting pollinators.
Neogene (Geol)	Geologic period of Late Tertiary age, alternatively Miocene + Pliocene.
neoteny	Supposed prolongation of larval or juvenile development stages into adult life.
nexine	Inner layer of exine; in general, can be regarded as synonym of endexine.

niche	Habitat available to be inhabited by a suitable organism.
Normapolles (Geol)	Prominent group of Late Cretaceous to Eocene triporate pollen, often with complex protruding apertures.
nucellus	Non-reproductive tissue of unfertilised ovule, the megasporangium in a seed-plant.
nymph	Early immature stage of an ectopterygote insect; see also 'larva'.
omnivore	Animal that feeds on both plants and animals.
organ-genus	Description of most fossil plant genera, i.e. representing only one organ; term no longer officially employed in ICBN.
Ornithischia	Group of dinosaurs with bird-like pelvic (hip) girdle.
ornithopod	Beaked dinosaur.
orthotropous	Of ovules that are upright and not inverted; cp. anatropous.
ostracod	Small aquatic arthropod with a bivalve calcareous carapace, commonly fossilised.
outcrop (Geol)	Part of rock body visible at the earth's surface.
ovary	Lower part of gynoecium containing ovules.
ovule	Megasporangium with single functional megaspore, enclosed by integument(s).
ovuliferous scales	Scales bearing ovules in extant conifer cones.
oxidation (Geol)	Here refers to critical process in paleopalynological preparation technique, nowadays kept to minimum duration.
palaeo- (Geol)	Spelling to reveal etymology, usually preferred in Europe for 'paleo' used in this work.
Paleogene (Geol)	Geologic period of Early Tertiary age, alternatively Paleocene, Eocene and Oligocene.
paleogeography (Geol)	Reconstruction of past geography for a specified period of time.
paleolatitude (Geol)	Past latitude of a rock formation and thus of a locality, estimated from remanent paleomagnetic properties of the rocks.
paleomagnetic (Geol)	Of (fossil) remanent magnetism of rocks, which can be measured.
paleopalynology (Geol)	Part of palynology concerned with all fossils; for convenience usually taken to exclude Quaternary occurrences.
paleosol (Geol)	Rock identified as representing a fossil soil horizon.
paleotaxon (Geol)	Immutable base-taxon of fossils, used in PDHC (see below); replaces (Hughes 1989) the concept of a biorecord (Hughes and Moody-Stuart 1969).
palmate	Of compound leaves with four or more laminae arising from a single point.
palynevent (Geol)	Estimate of nature of sedimentary environment, based on observations of palynomorphs and palynodebris.
palynoflora	Assemblage of all acid-resistant microfossils, obtained by preparation from a rock sample.
palynology	Study of all palynomorphs, fossil or extant.
palynomorph	Collective term for spores, pollen, dinocysts and other algae.

pandan	Member of the extant monocotyledonous Pandanaceae, pantropical coastal shrubs and small trees.
Pangaea (Geol)	Hypothetical pre-Triassic supercontinent of Laurasia and Gondwana, subsequently presumed to have fragmented.
papillate	Of blunt projections from cells of plant epidermis.
paracytic	Of stomata, with two subsidiary cells adjacent to two guard cells, and thus with bilateral symmetry; cp. syndetocheile.
paraphyletic	Of a group of organisms that does *not* contain all the descendants of a common ancestor
parasitic	Of an organism living entirely at the expense of another organism.
PDHC (Geol)	Paleontologic Data Handling Code (see Hughes 1989).
peltate	Of a leaf or ovule with a centrally attached petiole or stalk.
pentamerous	Of flower with floral parts arranged in fives.
perianth	Collective term for all elements of a flower below or outside the stamens.
period (Geol)	Large division of global geologic time-scale of duration between 10 and 100 Ma, e.g. Jurassic period.
Period Classification (Geol)	Method of classification of fossil plants according only to their general age of occurrence; preferred here to general hierarchical botanical classification, regardless of time.
periporate	Angiosperm pollen with pores equally distributed over the whole exine surface.
permineralised (Geol)	Of fossil plant material preserved by impregnation of mineral matter; this term now preferred in place of 'petrifaction' (see below).
petal	Member of innermost cycle of non-fertile appendages in a flower.
petiole	Stalk supporting a single leaf lamina.
petrifaction (Geol)	Plant fossil in which the original cell spaces are impregnated by mineral matter in diagenesis. See 'permineralised'.
phase	Term here employed for successive stages of development of diversity of fossil MCT (see above) pollen.
phloem	Tissue of sieve-tube elements formed outside the xylem and for conducting products of photosynthesis away from the leaves to other parts of the plant.
phylogeny	Evolutionary succession of organisms.
Pinaceae	Large family of extant conifers bearing needle-like leaves, in which resin (see amber) is secreted from the wood.
pinnate	Of compound leaves with a row of leaflets on either side of a single axis.
pistillate	Of flowers bearing carpels but not stamens.
plankton	Collective term for all organisms floating (i.e. subject to currents) in seas or lakes.
platanoid	Of Late Cretaceous plant fossils displaying one or more characters of the extant Platanaceae.
plate tectonics (Geol)	Method of interpretation large-scale horizontal movements of continents or of other regions of crustal rocks.
pollen	Microspore of seed plants containing the male gametophyte.

pollen chamber	Chamber formed above the apex of an ovule, enclosed by integument, for received pollen to germinate and fertilise the ovule.
Pollenifera (Geol)	New collective term without hierarchical rank, for all seed-plants in Mesozoic rocks; adopted in this work.
pollenkitt	Sticky substance observed on surface of many angiosperm pollen grains; probably concerned in insect pollination.
pollen tube	Tube that develops out of a pollen grain on germination, for fertilisation by passage of the male nucleus to the female organ.
pollination	Process of transferring pollen from its point of formation to a receptive surface of ovule or stigma (carpel).
polyphyletic	Of a group of organisms believed to have been derived in the period under discussion from several ancestors.
polyplicate	Of pollen with multiple longitudinal linear thickenings or folds of the exine, e.g. extant *Ephedra*.
polyploid	Plant with more than two sets of chromosomes (triploid for three and so on).
polyporate	Of angiosperm pollen with pores distributed over whole surface; effectively synonym of periporate.
pore	Circular thinning of exine in pollen grain, associated with germination.
preparation (Geol)	Extraction of palynomorphs from rock sample mounted on a glass microscope slide.
primitive	Extant plant with a low Advancement Index.
pro-angiosperm	Fossil plant with one or more angiosperm characters, believed to be a precursor of angiosperms.
progeny	Collective term for successful reproductive products of an organism.
propagule	Collective term for propagative plant organs, e.g. fruit, seeds and spores.
protandrous	Of plants in which pollen is ripened and dispersed, before the female ovular structure is receptive.
proximal	Of the pole of a spore or pollen grain adjacent to that of other members of the tetrad; cp. distal.
psilate	Of a pollen or spore surface observed under the microscope to be smooth.
pteridophyte	Vascular plants producing spores only; taken to include extant ferns, lycopods and horsetails.
pteridosperm	Group of seed-plants, originally based on Carboniferous fern-like plants that produced seeds; probably not a natural group.
punctuated equilibrium (Geol)	Supposed evolutionary pattern in which evolution proceeds rapidly in short bursts, with equilibrium between.
pupa	Distinct enclosed stage in Endopterygota between active larval stage and adult imago.
pycnoxylic	Of secondary wood of a dense regular nature comprising, almost entirely, tracheids.
pyritic (Geol)	Of a fossil preservation in which cellular spaces have been preserved by impregnation of iron pyrites (FeS_2).
pyroclastic (Geol)	Of fragmentary material thrown out of an active volcano.

radiation	Evolutionary pattern of very rapid appearance of many diverse descendants of one lineage.
radula	Toothed rasping organ characteristic of gastropod mollusca.
rainforest	Tropical forest in which dominant plants are trees; high rainfall.
Ranales	Old collective term for *Ranunculus* group, also applied to all Magnoliidae; can cause confusion.
rattan	Scrambling, very spiny, member of the monocotyledonous Palm family.
receptacle	Part of a floral axis on which reproductive structures are borne.
rendzina	Shallow intrazonal soil rich in lime.
reticulate	Of sculpture of spores and pollen, of more or less regular network of ridges (muri) with spaces between (lumina).
reworking (Geol)	In some circumstances microfossils may be eroded and redeposited in later strata; detection of this is often possible.
Rosidae	Subclass of angiosperms with mainly actinomorphic flowers, e.g. rose, saxifrage.
r-selected	Of the reproductive 'strategy' of species characteristic of unstable environments, reproducing quickly with abundant progeny.
saccate	Of pollen with one or more sacci.
salinity	Degree of content of various salts in (sea) water.
samara	Winged fruit.
sample (Geol)	Material collected at one time from a locality (see above) and stored.
saprophyte	Plant drawing sustenance from dead material of other plants, e.g. many fungi.
Saurischia	Lizard-hipped dinosaurs; cp. Ornithischia.
sauropod	Suborder of large quadripedal dinosaurs.
savannah	Region of tropical, predominantly grass, vegetation with some scattered trees.
sculpture	External textural projections or indentations on the surface of a spore or pollen grain.
sea-level	Mean level at any time by reference to a standard.
seed	Mature fertilised ovule containing an embryo.
seed-fern	See 'pteridosperm' (above).
SEM	Scanning electron microscope.
SEMG	Scanning electron micrograph.
semi-tectate	Of tectate pollen, of which the tectum is reticulate or otherwise incomplete.
sepal	Sterile outer part of flower perianth, below the petals.
shoot	Vascular plant axis, mainly equivalent to 'stem', which is the part above ground.
soil	Product of weathering of rocks *in situ*, and of organisms; many local types.
species	Traditional base-taxon composed of living individuals capable of interbreeding; 'species' of fossils cannot be so based.

'splitting'	Style of taxonomy of fossils in which small differences are recorded in new taxa; see also 'lumping'.
sporangium	Organ in pteridophytes in which spores are borne.
spore	Disseminule of pteridophytes and bryophytes, normally with trilete or monolete proximal laesura.
sporophyte	Spore-producing asexual diploid ($2n$) stage of plant alternation of generations.
sporopollenin	Resistant polymer substance of spore and pollen exine; surprisingly durable through geologic time through all but very high temperatures.
stamen	Pollen-producing organ in a flower.
stigma	Receptive distal part of carpel, where pollen alights and germinates.
stoma	Pore in the plant epidermis surrounded by guard cells; plural stomata.
stratigraphic correlation (Geol)	Correlation of time of formation of rock successions of approximately similar age separated in space.
stratigraphic scale (Geol)	See time-scale (below).
stratigraphy (Geol)	Study of succession of formation of crustal rocks, from positional and petrologic relationships between strata.
strew-search (Geol)	Search for dispersed microfossils on gridded SEM (see above) stub, by traversing as with a light microscope.
striate	Of pollen with multiple parallel grooves and ridges; term better retained for fine striae, excluding taeniate and polyplicate.
structure	Features of arrangement of layers and other elements within the pollen wall.
stub	Standard metal target for mounting specimens for SEM (see above) observation.
stub-grid	A suitable metal grid employed on the surface of a scanning electron microscope stub for locating individual small specimens.
style	Gynoecial extension on which a raised stigma is placed.
sulcus	Elongate furrow (aperture) across distal pole of some pollen grains.
supra-mural	Of pollen exine sculpture above or outside the surface of the tectum.
synangia	Groups of united sporangia.
syncarpous	Of angiosperm gynoecia with carpels fused together.
syndetocheile	Of stomata, in which guard cells and subsidiary cells develop from the same initial; cp. haplocheile.
taeniate	Of pollen exine sculpture consisting of flat raised straps of ectoexine.
taphonomy (Geol)	Study of the effect of details of burial and diagenesis on the mode of preservation of fossils.
taxon	Formal group of organisms, believed to have resemblances leading to useful classification, e.g. species, genus, family, etc.
tectate	Of pollen grains in which a tectum is recognised.
tectum	Outer layer of ectexine of pollen grains.
TEM	Transmission electron microscope.
tepal	Individual flower corolla part, when petals and sepals are not distinguished as in some monocotyledons.

terrestrial	Of organisms living on land.
terrigenous (Geol)	Sediments directly derived from erosion of land.
Tethys (Geol)	Sea presumed to have separated Eurasia and Africa in Mesozoic and Paleogene times.
tetrad	Symmetrical group of four pollen grains or spores, resulting from the meiotic division of one mother cell.
time-correlation (Geol)	Frequently used to mean succession correlation because successions themselves indicate passage of time.
time-scale (Geol)	Scale of successive specified divisions of time based on marker-points in stratified rock.
topotype	Specimen of a fossil taken from the sample from which a base-taxon has previously been characterised.
tracheid	Xylem water-conducting elongated lignified cell, without further elaboration; characteristic of extant pteridophytes and gymnosperms.
transgression (Geol)	Spread of sea over land area, detectable from changed stratification, regardless of tectonic or other cause.
trash	Term applied to all plant material, (fallen and) no longer living.
tree-ferns	Ferns of extant families Cyatheaceae and Dicksoniaceae with false trunks of up to 10 m; several fossil analogues.
triaperturate	Collective term for angiosperm pollen of tricolpate, triporate, tricolporate and other types.
trichotomosulcate	Of monosulcate pollen grains, in which the (distal) sulcus is three-rayed, simulating but not representing a trilete laesura (proximal).
tricolpate	Of pollen with three simple meridionally arranged (120° apart) colpi; characteristic of extant dicotyledonous angiosperms; first fossil records in Mid-Cretaceous rocks.
trimerous	Of flowers with floral parts arranged in threes; e.g. monocotyledons.
tripinnate	Thrice divided pinna, providing a large complex leaf as in some pteridosperms.
triporate	Of pollen grains with three pores 120° apart, arranged equatorially.
tropical	Of climates experienced between the tropics, perhaps distinctive mainly in glacial periods (as now).
ulcerate	Of pollen with irregular thinning of a distal pore.
unconformity (Geol)	Break of kind or of inclination between adjacent superposed strata, which indicates a hiatus of non-deposition.
unconsolidated (Geol)	Of rocks consisting of loose sediment unaffected by compression and diagenesis.
uniformitarian (Geol)	Of the view that past geologic processes have remained the same through time; a misleading simplification.
unisexual	Of flowers or plant individuals displaying only stamens or only carpels.
upland (Geol)	Land areas of erosion, above and distinct from aggradational land.
valves	Of two halves of a capsule split open to reveal ovules or seeds; as in fossil *Leptostrobus*.
vessel	Longitudinal tubes placed end to end in xylem of most angiosperms.
volcaniclastic (Geol)	Sediment directly formed from volcanic fragmentary deposits.

Glossary

water-ferns	Extant heterosporous ferns that grow floating in water or near water, e.g. *Marsilia, Salvinia, Azolla*.
xeromorphic	Of plants with characters such as thick cuticles; typical of dry habitats.
xerophytic	Of plants of dry habitats.
xylem	Water-conducting tissue of vascular plants, consisting of tracheids but also with vessels in angiosperms.
zone (Geol)	Generally understood as a unit of stratified rock characterised by a geologic phenomenon, of which the name is used as a prefix to -zone.
zone-fossil (Geol)	Fossil employed to characterise a zone (see above).
zonisulcate	Of pollen with a ring furrow equatorially or just distal of that, e.g. *Classopollis*.
zygomorphic	Of flowers with bilateral symmetry.

References

Allen, P. and Wimbledon, W. A. (1991). Correlation of NW European Purbeck–Wealden (nonmarine Lower Cretaceous) as seen from the English type-areas. *Cret. Res.* **12**, 511–26.

Alvin, K. L. (1957). On *Pseudaraucaria* Fliche emend., a genus of fossil pinaceous cones. *Ann Bot.* (NS) **21**, 33–51.

Alvin, K. L. (1977). The conifers *Frenelopsis* and *Manica* in the Cretaceous of Portugal. *Palaeontology* **20**, 387–404.

Alvin, K. L. (1982). Cheirolepidiaceae: biology, structure and palaeoecology. *Rev. Palaeobot. Palynol.* **37**, 71–98.

Alvin, K. L. and Chaloner, W. G. (1970). Parallel evolution in leaf venation: an alternative view of angiosperm origins. *Nature* **226**, 662–3.

Alvin, K. L., Spicer, R. A. and Watson, J. (1978). A *Classopollis*-containing male cone associated with *Pseudofrenelopsis*. *Palaeontology* **21**, 847–56.

Anderson, J. M. and Anderson, H. M. (1983). Vascular plants from the Devonian to Lower Cretaceous in Southern Africa. *Bothalia* **3/4**, 337–44.

Anderson, J. M. and Anderson, H. M. (1985). *Prodromus of South African megafloras Devonian to Lower Cretaceous*. Balkema, Rotterdam, 423 pp.

Anderson, J. M. and Anderson, H. M. (1989). *Palaeoflora of Southern Africa: Molteno Formation (Triassic)*; Vol. 2, *Gymnosperms (excluding* Dicroidium). Balkema, Rotterdam, pp. 1–567.

Andrews, H. N. (1980). *The fossil hunters in search of ancient plants*. Cornell University Press, Ithaca, NY, 421 pp.

Andrews, H. N. and Kern, E. M. (1947). The Idaho Tempskyas and associated fossil plants. *Ann. MO Bot. Gard.* **34**, 119–86.

Arber, E. A. N. and Parkin, J. (1907). On the origin of angiosperms. *Bot. J. Linn. Soc.* **38**, 29–80.

Archangelsky, S. (1965). Fossil Ginkgoales from the Ticó flora, Santa Cruz Province, Argentina. *Bull. Brit. Mus. (Nat. Hist.) Geol.* **10**, 119–37.

Archangelsky, S. (1968*a*). Studies on Triassic fossil plants from Argentina. IV. The leaf genus *Dicroidium* and its possible relation to *Rhexoxylon* stems. *Palaeontology* **11**, 500–12.

Archangelsky, S. (1968*b*). On the genus *Tomaxiella* (Coniferae). *Bot. J. Linn. Soc.* **61**, 153–65.

Archangelsky, S. and Taylor, T. N. (1993). The ultrastructure of in situ *Clavatipollenites* pollen from the Early Cretaceous of Patagonia. *Am. J. Bot.* **80**, 879–85.

Arkell, W. J. (1947). The geology of the country around Weymouth, Swanage, Corfe and Lulworth. *Mem. Geol. Surv. G.B.* 1–386.

Arkell, W. J. (1956). *Jurassic geology of the world.* Oliver and Boyd, London, 806 pp.

Arnold, C. A. (1948). Classification of gymnosperms from the viewpoint of paleobotany. *Bot. Gaz.* **110**, 2–12.

Ash, S. R. (1976). Occurrence of the controversial plant fossil *Sanmiguelia* in the Upper Triassic of Texas. *J. Paleontol.* **50**, 799–804.

Ash, S. R. and Read, C. B. (1976). North American species of *Tempskya* and their stratigraphic significance. *US Geol. Surv. Prof. Paper* **874**, 1–42.

Bakker, R. T. (1987). *The dinosaur heresies.* Longman, UK, 480 pp.

Barale, G. (1981). La paléoflore Jurassique du Jura français. *Docum. Lab. Géol. Lyon* **81**, 1–467.

Barron, E. J. (1983). A warm, equable Cretaceous; the nature of the problem. *Earth Sci. Rev.* **19**, 305–18.

Batten, D. J. (1973). Use of palynologic assemblage-types in Wealden correlation. *Palaeontology* **16**, 1–40.

Beard, J. S. (1989). The early evolution of plant life in Western Australia. *J. R. Soc. West. Austr.* **71**, 59–67.

Berry, E. W. (1916). A petrified palm from the Cretaceous of New Jersey. *Am. J. Sci.* (IV) **41**, 193–7.

Bessey, C. E. (1915). Angiosperm phylogeny. *Ann. MO. Bot. Gard.* **2**, 109.

Bose, M. N. and Manum, S. B. (1990). Mesozoic conifer leaves with 'Sciadopitys-like' stomatal distribution. A re-evaluation based on fossils from Spitsbergen, Greenland and Baffin Island. *Skr. Nor. Polarinstit., Oslo* **192**, 81 pp.

Bose, M. N., Pal, P. K. and Harrris, T. M. (1985). The Pentoxylon plant. *Philos. Trans. R. Soc. Lond., ser. B.* **310**, 77–108.

Brenner, G. J. (1963). The spores and pollen of the Potomac Group of Maryland. *Md Dept. Geol. Mines Water Resour. Bull.* **27**, 1–215.

Brenner, G. J. (1967). The gymnospermous affinity of *Eucommiidites* Erdtman 1948. *Rev. Palaeobot. Palynol.* **5**, 123–7.

Brenner, G. J. (1976). Middle Cretaceous floral provinces and early migration of angiosperms. In Beck, C. B. (Ed.) *Origin and early evolution of angiosperms.* Columbia University Press, New York, pp. 23–47.

Brenner, G. J. (1984). Late Hauterivian angiosperm pollen from the Helez Formation, Israel. *Sixth Int. Palynol. Conf. Calgary, Abstracts*, 15.

Brenner, G. J. and Bickoff, I. S. (1992). Palynology and age of the Lower Cretaceous Basal Kurnub Group from the coastal plain to the Northern Negev of Israel. *Palynology* **16**, 137–85.

Brongniart, A. (1822). Sur la classification et la distribution des végétaux fossiles en général, et sur ceux des terrains de sédiment supérieur en particulier. *Mus. Hist. Nat. (Paris) Mém.* **8**, 203–348.

Brown, R. W. (1956). Palm-life plants from the Dolores Formation (Triassic), southwestern Colorado. *US Geol. Surv. Prof. Paper* **274H**, 205–9.

Buckland, W. (1828). On the Cycadeoideae, a family of fossil plants found in the Oolite quarries of the Isle of Portland. *Trans. Geol. Soc. ser.* ii, **2**, 395–401.

Burger, D. (1980). Palynology of the Lower Cretaceous in the Surat Basin. *Bull. Bur. Miner. Resour. Geol. Geophys. Aust.* **189**, 1–106.

Burger, D. (1993). Early and middle Cretaceous angiosperm pollen grains from Australia. *Rev. Palaeobot. Palynol.* **78**, 183–234.

Burger, W. C. (1981a). Heresy revived: the monocot theory of angiosperm origin. *Evol. Theory* **5**, 189–225.

Burger, W. C. (1981*b*). Why are there so many kinds of flowering plant? *Bioscience* **31**, 572–81.

Carpenter, F. M. (1992). *Treatise on invertebrate paleontology*, Part R, *Arthropoda*, Volumes 3 and 4, *Superclass Hexapoda*. Geological Society of America, Washington, p. xxii and 655 pp.

Chaloner, W. G. (1970). The evolution of miospore polarity. *Geosci. Man* **1**, 47–56.

Chaloner, W. G. (1985). Thomas Maxwell Harris (1903–1983) *Biograph. Mem. Fellows R. Soc.* **31**, 229–60.

Chaloner, W. G., Friis, E. M. and Hemsley, A. R. (1991). *Cycadocephalus* Nathorst, a fern not a bennettite. *Neues Jahr. Geol. Paläontol. Abh. Stuttgart* **183**, 347–62.

Chaloner, W. G. and Hemsley, A. R. (1991). Heterospory: cul-de-sac or pathway to the seed. In Blackmore, S. and Barnes, S. H. (Eds.) *Pollen and spores. Syst. Assoc. Spec.* Vol. **44**, 151–167.

Chandler, M. E. J. (1964). *The Lower Tertiary floras of Southern England, IV*. British Museum (Nat. Hist.), London.

Chandler, M. E. J. (1978). Supplement to the Lower Tertiary floras of Southern England, V. *Tert. Res. Spec. Papers* **4**, 1–47.

Chapman, J. L. (1987*a*). Sporne's Advancement Index re-visited. *New Phytol.* **106**, 319–32.

Chapman, J. L. (1987*b*). Comparison of Chloranthaceae pollen with the Cretaceous '*Clavatipollenites* complex'; taxonomic implications for palaeopalynology. *Pollen Spores* **29**, 249–72.

Clement-Westerhof, J. A. and Van Konijnenburg-van Cittert, J. H. A. (1991). New data on the fertile organs leading to a revised concept of the Cheirolepidiaceae. *Rev. Palaeobot. Palynol.* **68**, 147–79.

Coe, M. J., Dilcher, D. L., Farlow, J. D., Jarzen, D. M. and Russell, D. A. (1987). Dinosaurs and land plants. In Friis, E. M., Chaloner, W. G. and Crane, P. R. (Eds.). *The origins of angiosperms and their biological consequences.* Cambridge University Press, Cambridge, pp. 225–58.

Collinson, M. E. (1982). A preliminary report on the Senckenberg-Museum collection of fruits and seeds from Messel bei Darmstadt. *Courier Forschungsinstitut Senckenberg* **56**, 49–57.

Collinson, M. E. (1983). *The London Clay Flora.* Field Guides to Fossils, No. 1, The Palaeontological Association, London, 121 pp.

Collinson, M. E. and Hooker, J. J. (1987). Vegetational and mammalian faunal changes in the Early Tertiary of Southern England. In Friis, E. M., Chaloner, W. G. and Crane, P. R. (Eds.). *The origins of angiosperms and their biological consequences.* Cambridge University Press, Cambridge, pp. 259–304.

Cope, J. C. W., Duff, K. L., Parsons, C. F., Torrens, H. S., Wimbledon, W. A. and Wright J. K. (1980). A correlation of Jurassic rocks in the British Isles, Part 2, Middle and Upper Jurassic. *Geol. Soc. Lond. Spec. Rep.* **15**, 109 pp.

Corner, E. J. H. (1966). *The natural history of palms.* Weidenfeld and Nicolson, London, 393 pp.

Cornet, B. (1977). The palynology and age of the Newark Supergroup. Ph.D. thesis, Pennsylvania State University.

Cornet, B. (1986). The reproduction structures and leaf venation of a Late Triassic angiosperm *Sanmiguelia lewisii. Evol. Theory* **7**, 231–309.

Cornet, B. (1989*a*). Late Triassic angiosperm-like pollen from the Richmond rift basin of Virginia, USA. *Palaeontographica* **213 B**, 37–87.

Cornet, B. (1989*b*). The reproduction morphology and biology of *Sanmiguelia lewisii*, and its bearing on angiosperm evolution in the Late Triassic. *Evol. Trends Plants* **3**, 25–51.

Cornet, B. and Habib, D. (1992). Angiosperm-like pollen flora the ammonite-dated Oxfordian (Upper Jurassic) of France. *Rev. Palaeobot. Palynol.* **71**, 269–94.

Cornford, F. M. (1908). *Microcosmographica academica.* Bowes, Cambridge.

Couper, R. A. (1953). Upper Mesozoic and Cainozoic spores and pollen grains from New Zealand. *NZ Geol. Surv. Palaeontol. Bull.* **22**, 1–77.

Couper, R. A. (1958). British mesozoic microspores and pollen grains, a systematic and stratigraphic study. *Palaeontographica* **103 B**, 75–179.

Cowie, J. W., Ziegler, W., Boucot, A. J., Bassett, M. G. and Remane, J. (1986). Guidelines and statutes of the International Commission on Stratigraphy (ICS). *Courier Forschungsinstitut Senckenberg* **83**, 1–14.

Crane, P. R. (1984). Misplaced pessimism and misguided optimism: a reply to Mabberley. *Taxon* **33**, 79–82.

Crane, P. R. (1985). Phylogenetic analysis of seed plants and the origin of angiosperms. *Ann. MO Bot. Gard.* **72**, 716–93.

Crane, P. R. (1987). Vegetational consequences of the angiosperm diversification. In Friis, Chaloner, W. G. and Crane, P. R. (Eds.) *The origin of angiosperms and their biological consequences.* Cambridge University Press, Cambridge, pp. 105–44.

Crane, P. R. (1988). Major clades and relationships in the 'Higher' Gymnosperms. In Beck, C. B. (Ed.) *Origin and evolution of gymnosperms.* Columbia University Press, New York, pp. 218–72.

Crane, P. R. and Dilcher, D. L. (1984). *Lesqueria*: an angiosperm fruiting axis from the Mid-Cretaceous. *Ann. MO Bot. Gard.* **71**, 384–402.

Crane, P. R., Friis, E. M. and Pedersen, K. R. (1986). Lower Cretaceous angiosperm flowers: fossil evidence on early radiation of dicotyledons. *Science* **232**, 852–4.

Crane, P. R., Friis, E. M. and Pedersen, K. R. (1989). Reproductive structure and function in Cretaceous Chloranthaceae. *Plant Syst. and Evol.* **165**, 211–26.

Crane, P. R. and Upchurch, G. R. Jr (1987). *Drewria potomacensis* gen. et sp. nov., an Early Cretaceous member of the Gnetales from the Potomac Group of Virginia. *Am. J. Bot.* **74**, 1722–36.

Crepet, W. L. (1974). Investigations of North American cycadeoids: the reproductive biology of *Cycadeoidea. Palaeontographica* **148 B**, 144–69.

Crepet, W. L. and Friis, E. M. (1987). The evolution of insect pollination in angiosperms. In Friis, E. M., Chaloner, W. G. and Crane, P. R. (Eds.). *The origins of angiosperms and their biological consequences.* Cambridge University Press, Cambridge, pp. 181–201.

Crepet, W. L., Friis, E. M. and Nixon, K. C. (1991). Fossil evidence for the evolution of biotic pollination. *Philos. Trans. R. Soc. Lond.* **333**, 187–95.

Crepet, W. L., Nixon, K. C., Friis, E. M. and Freudenstein, J. V. (1992). Oldest fossil flowers of hamamelidaceous affinity, from the Late Cretaceous of New Jersey. *Proc. Natl. Acad. Sci. USA* **89**, 8986–9.

Cridland, A. A. (1957). *Williamsoniella papillosa*, a new species of bennettitalean flower. *Ann. Mag. Nat. Hist. ser 10* **12**, 383–8.

Cronquist, A. (1968). *The evolution and classification of flowering plants*. Nelson, London, 396 pp.

Cuerrier, A., Barabe, D. and Brouillet, L. (1992). Bessey and Engler – a numerical analysis of their classification of the flowering plants. *Taxon* **41**, 667–684.

Daghlian, C. P. (1981). A review of the fossil record of monocotyledons. *Bot. Rev.* **47**, 517–55.

Darwin, C. 1859. *On the origin of species by natural selection, or the preservation of favoured races in the struggle for life*. John Murray, London.

Davies, D. (1929). Correlations and palaeontology of the Coal Measures in eastern Glamorganshire. *Phil. Trans. R. Soc. Lond.* **217**.

Davis, P. H. and Heywood, V. H. (1963). *Principles of angiosperm taxonomy*. Oliver and Boyd, Edinburgh and London, 556 pp.

Delevoryas, T. (1968). Some aspects of cycadeoid evolution. *Bot. J. Linn Soc.* **61**, 137–46.

Delevoryas, T. and Gould, R. E. (1973). Investigation of North American cycadeoids: Williamsonian cones from the Jurassic of Oaxaca, Mexico. *Rev. Palaeobot. Palynol.* **15**, 27–42.

Dettmann, M. E. (1986). Early Cretaceous palynoflora of subsurface strata correlative with the Koonwarra Fossil Bed, Victoria. In Jell, P. A. and Roberts, J. (Eds.) *Plants and invertebrates from the Lower Cretaceous Koonwarra Fossil Bed, South Gippsland, Victoria. Mem. Assoc. Austral. Palaeontol.* **3**, 79–110.

Dettmann, M. E. and Clifford, H. T. (1991). Spore morphology of *Anemia, Mohria* and *Ceratopteris* (Filicales). *Am. J. Bot.* **78**, 303–25.

Dettman, M. E. and Clifford, H. T. (1992). Phylogeny and biogeography of *Ruffordia, Mohria* and *Anemia* (Schizaeaceae) and *Ceratopteris* (Pteridaceae): evidence from *in situ* and dispersed spores. *Alcheringa* **16**, 269–314.

Dettmann, M. E., Molnar, R. E., Douglas, J. G., Burger, D., Fielding, C., Clifford, H. T., Francis, J., Jell, P., Rich, T. H. V., Wade, M., Rich, P. V., Pledge, N., Kemp, A., Rozefelds, A. (1992). Australian Cretaceous terrestrial faunas and floras; biostratigraphic and biogeographic implications. *Cret. Res.* **13**, 207–62.

Dijkstra, S. J. (1949). Megaspores etc., from the Senonian of the South Limburg. *Meded. geol. Sticht.*, s'Gravenhage, **NS3**, 19–32.

Dijkstra, S. J. (1951). Wealden megaspores and their stratigraphical value. *Meded. geol. Sticht.*, s'Gravenhage, **NS5**, 7–21.

Dilcher, D. L. (1979). Early angiosperm reproduction: an introductory report. *Rev. Palaeobot. Palynol.* **27**, 291–328.

Dilcher, D. L. and Crane, P. R. (1984). *Archaeanthus*: an early angiosperm from the Cenomanian of the Western Interior of North America. *Ann. MO Bot. Gard.* **71**, 351–83.

Dilcher, D. L. and Kovach, W. L. (1986). Early angiosperm reproduction: *Caloda delevoryana* gen. et sp. nov., a new fructification from the Dakota Formation (Cenomanian) of Kansas. *Am. J. Bot.* **73**, 1230–7.

Dodson, P., Behrensmeyer, A. K., Bakker, R. T. and McIntosh, J. S. (1980). Taphonomy and paleoecology of the dinosaur beds of the Jurassic Morrison Formation. *Paleobiology* **6**, 208–32.

Donoghue, M. J. and Doyle, J. A. (1989). Phylogenetic analyses of angiosperms and the relationships of Hamamelidae. In Crane, P. R. and Blackmore, S. (Eds.) *Evolution, systematics and fossil history of the Hamamelidae*, Vol. 1, *Introduction and 'Lower' (Hamamelidae). Syst. Assoc. Spec. Vol.* **40A**, 17–45.

Donoghue, M. J. and Doyle, J. A. (1991). Angiosperm monophyly. *Tree*, **6**, 407.

Donoghue, M. J., Doyle, J. A., Gauthier, J., Kluge, A. G. and Rowe, T. (1989). The importance of fossils in phylogeny reconstruction. *Ann. Rev. Ecol. Syst.* **20**, 431–60.

Douglas, J. G. (1969). The Mesozoic floras of Victoria, Parts 1 and 2. *Geol. Surv. Victoria Mem.* **28**, 310 pp.

Douglas, J. G. (1973). The Mesozoic floras of Victoria, Part 3. *Geol. Surv. Victoria Mem.* **29**, 185 pp.

Doyle, J. A. (1973). The monocotyledons: their evolution and comparative biology. V. Fossil evidence on early evolution of the monocotyledons. *Q. Rev. Biol.* **48**, 399–413.

Doyle, J. A. (1976). Man bites botanical dogma. *Paleobiology* **2**, 265–78.

Doyle, J. A. (1978*a*). Fossil evidence on the evolutionary origin of tropical trees and forests. In Tomlinson, P. B. and Zimmermann, M. H. (Eds.) *Tropical trees as living systems*. Cambridge University Press, Cambridge, pp. 3–30.

Doyle, J. A. (1978*b*). On the origin of angiosperms. *Ann. Rev. Ecol. Syst.* **9**, 365–92.

Doyle, J. A. (1982). Palynology of Continental Cretaceous sediments, Crisfield Geothermal Test Well, Eastern Maryland. *MD Geol. Surv. Open File Rep. Wastgate Formation*, 51–87.

Doyle, J. A. (1983). Palynological evidence for Berriasian age of basal Potomac Group sediments, Crisfield Well, Eastern Maryland. *Pollen Spores* **25**, 499–530.

Doyle, J. A. (1984). Evolutionary, geographic and ecological aspects of the rise of angiosperms. *Proc. 27th Int. Bot. Congr.* Vol. 2, Palaeontology, pp. 23–33.

Doyle, J. A. (1988). Pollen evolution in seed plants: a cladistic perspective. *J. Palynology* **23–24**, 7–18.

Doyle, J. A. (1992). Revised palynological correlation of the Lower Potomac Group (USA) and the Cocobeach sequence of Gabon (Barremian–Aptian). *Cret. Res.* **13**, 337–49.

Doyle, J. A., Biens, P., Doerenkamp, A. and Jardine, S. (1977). Angiosperm pollen from the pre-Albian Lower Cretaceous of Equatorial Africa. *Bull. Centres Rech. Explor.-Prod. Elf-Aquitaine* **1**, 451–73, 2 pls.

Doyle, J. A. and Donoghue, M. J. (1987). The importance of fossils in elucidating seed plant phylogeny and macroevolution. *Rev. Palaeobot. Palynol,* **50**, 63–95.

Doyle, J. A. and Donoghue, M. J. (1992). Fossil and seed plant phylogeny re-analysed. *Brittonia* **44**, 89–106.

Doyle, J. A. and Donoghue, M. J. (1993). Phylogenies and angiosperm diversification. *Paleobiology* **19**, 141–67.

Doyle, J. A. and Hickey, L. J. (1976). Pollen and leaves from the Mid-Cretaceous Potomac Group and their bearing on early angiosperm evolution. In Beck, C. B. (Ed.) *Origin and early evolution of angiosperms*. Columbia University Press, New York, pp. 139–206.

Doyle, J. A. and Hotton, C. L. (1991). Diversification of early angiosperm pollen in a cladistic context. In Blackmore, S. and Barnes, S. H. (Eds.) *Pollen and spores, patterns of diversification. Syst. Assoc. Spec. Vol.* **44**, 169–95.

Doyle, J. A., Hotton, C. L. and Ward, J. V. (1990a). Early Cretaceous tetrads, zonosulc-ate pollen, and Winteraceae. I. Taxonomy, morphology and ultrastructure. *Am. J. Bot.* 77, 1544–57.

Doyle, J. A., Hotton, C. L. and Ward, J. V. (1990b). Early Cretaceous tetrads, zonosulc-ate pollen. II. Cladistic analysis and implications. *Am. J. Bot.* 77, 1558–68.

Doyle, J. A., Jardine, S., and Doerenkamp, A. (1982). *Afropollis*, a new genus of early angiosperm pollen, with notes on the Cretaceous palynostratigraphy and paleoenvironments of Northern Gondwana. *Bull. Centres Rech. Explor.-Prod. Elf-Aquitaine* 6, 39–117, 13 pls.

Doyle, J. A. and Robbins, E. I. (1977). Angiosperm pollen zonation of the continental Cretaceous of the Atlantic coastal plain and its application to deep wells in the Salisbury embrayment. *Palynology* 1, 43–78.

Drinnan, A. N. and Chambers, T. C. (1986). Flora of the Lower Cretaceous Koonwarra Fossil Bed (Konemburra Group), South Gippsland, Victoria. *Mem. Assoc. Austral. Palaeontol.* 3, 1–77.

Drinnan, A. N., Crane, P. R., Friis, E. M. and Pedersen, K. R. (1990). Lauraceous flowers from the Potomac Group (Mid-Cretaceous) of Eastern North Amer-ica. *Bot. Gaz.* 151, 370–84.

Drinnan, A. N., Crane, P. R., Friis, E. M. and Pedersen, K. R. (1991). Angiosperm flowers and tricolpate pollen of Buxaceous affinity from the Potomac Group (Mid-Cretaceous) of Eastern North America. *Am. J. Bot.* 78, 153–76.

Edwards, W. N. (1921). On a small bennettitalean flower from the Wealden of Sussex. *Ann. Mag. Nat. Hist. ser.* 9, 8, 440–2, 1 pl.

Endress, P. K. (1987). The Chloranthaceae: reproductive structure and phylogenetic position. *Bot. Jahrb. syst.* 109, 153–226.

Engler, A. and Prantl, H. 1897–1915. *Die naturlichen pflanzenfamilien*, 20 Vols. Leipzig.

Erdtman, G. (1948). Did dicotyledonous plants exist in Jurassic time? *Geol. Fören. Stockholm Förh.* 70, 265–71.

Erdtman, G. (1952). *Pollen morphology and plant taxonomy.* I. *Angiosperms.* Alm-quist and Wicksell, Stockholm, 539 pp.

Erwin, D. M. and Stockey, R. A. (1991). Silicified monocotyledons from the Middle Eocene Princeton Chart (Allenby Formation) of British Columbia, Canada. *Rev. Palaeobot. Palynol.* 70, 147–62.

Fisher, H. L. and Watson, J. (1983). A new conifer species from the Wealden beds of Feron-Glageon, France. *Bull. Br. Mus. (Nat. Hist.) Geol.* 37(3), 99–104.

Fitton, W. H. (1836). Observations on some of the strata between the Chalk and the Oxford Oolite in the South-East of England. *Trans. Geol. Soc. ser.* ii, 4, 103.

Frakes, L. A., Francis, J. E. and Syktus, J. L. (1992). *Climate modes of the Phanerozoic.* Cambridge University Press, Cambridge, 274 pp.

Francis, J. E. (1983). The dominant conifer of the Jurassic Purbeck Formation, England. *Palaeontology* 26, 277–94.

Francis, J. E. (1986). The calcareous paleosols of the Basal Purbeck Formation (Upper Jurassic), Southern England. In Wright, V. P. (Ed.) *Paleosols: their recogni-tion and interpretation.* Blackwell, Oxford, pp. 112–38.

Friis, E. M., Chaloner, W. G. and Crane, P. R. (Eds.) (1987). *The origins angiosperms and their biological consequences.* Cambridge University Press, Cambridge, 358 pp.

References

Friis, E. M. and Crane, P. R. (1989). Reproductive structures of Cretaceous Hama-melidae. In Crane, P. R. and Blackmore S.; *Evolution, systematics and fossil history of the Hamamelidae*, Vol. I. *Syst. Assoc. Spec. Vol.* **40A**, 155–74.

Friis, E. M., Crane, P. R. and Pedersen, K. R. (1988). Reproductive structures of Cretaceous Platanaceae. *Biol. Skri.* Copenhagen, **31**, 1–55.

Friis, E. M., Crane, P. R. and Pedersen, K. R. (1991). Stamen diversity and *in situ* pollen of Cretaceous angiosperms. In Blackmore, S. and Barnes, S. H. (Eds.) *Pollen and spores, patterns of diversification. Syst. Assoc. Spec. Vol.* **44**, 197–224.

Friis, E. M. and Crepet, W. L. (1987). Time of appearance of floral features. In Friis, E. M., Chaloner, W. G. and Crane, P. R. (Eds.) *The origins of angiosperms and their biological consequences.* Cambridge University Press, Cambridge, pp. 145–79.

Friis, E. M., Pedersen K. R. and Crane, P. R. (1992). *Esgueiria* gen. nov., fossil flowers with combretaceous features from the Late Creataceous of Portugal. *Biol. Skv. Dan. Vid. Selsk.*, Copenhagen, **41**, 5–45.

Friis, E. M. and Skarby, A. (1981). Structurally preserved angiosperm flowers from the Upper Cretaceous of Southern Sweden. *Nature*, **291**, 485–6.

Friis, E. M. and Skarby, A. (1982). *Scandianthus* gen. nov., angiosperm flowers of saxifragalean affinity from the Upper Cretaceous of Southern Sweden. *Ann. Bot.* **50**, 569–83.

Gallois, R. W. (1971). Log of Gayton Borehole. *Ann. Rep. Inst. Geol. Sci. UK* 1970, 23.

Gallois, R. W. (1972). Logs of Hunstanton and Skegness Boreholes. *Ann. Rep. Inst. Geol. Sci. UK* 1971, 116.

Gallois, R. W. (1984). The late Jurassic to mid Cretaceous rocks of Norfolk. *Bull. Geol. Soc. Norfolk*, **34**, 3–64.

Giraud, B., Bussert, R. and Schrank, E. (1992). A new Theacean wood from the Cretaceous of northern Sudan. *Rev. Palaeobot. Palynol.* **75**, 289–99.

Golenberg, E. M., Giannasi, D. E., Clegg, M. T., Smiley, C. J., Durbin, M., Henderson, D. and Zurawski, G. (1990). Chloroplast DNA sequence from a Miocene Magnolia species. *Nature* **344**, 656–8.

Goubeli, A. A., Hochuli, P. A. and Wildi, W. (1988). Lower Cretaceous turbidite sediments from the Rif Chain (northern Morocco). Palynology, stratigraphy and palaeogeographic setting. *Geol. Rundschau* **73**, 1081–114.

Greenwood, D. R. (1992). Taphonomic constraints on foliar physiognomic interpretations of Late Cretaceous and Tertiary palaeoclimates. *Rev. Palaeobot. Palynol.* **71**, 149–90.

Greuter, W. and McNeill, J. (Eds.) (1988). International Code of Botanical Nomenclature. *Regnum Vegetabile*, Vol. 118, Koeltz Scientific Books, Koenigstein, p. xiv, 328 pp.

Haeckel, E. (1866). *Generelle Morphologie der Organismen.* Berlin.

Harding, I. C. (1988). A remarkable new Early Cretaceous trilete spore with observations on sporoderm function and stratigraphic significance. *Rev. Palaeobot. Palynol.* **54**, 165–73.

Harding, I. C. (1990). A dinocyst calibration of the European Boreal Barremian. *Palaeontographica* **218 B**, 1–76.

Harland, W. B. (1992). Stratigraphic regulation and guidance: a critique of current tendencies in stratigraphic codes and guides. *Geol. Soc. Am. Bull.* **104**, 1231–5.

Harland, W. B., Armstrong, R. L., Cox, A. V., Craig, L. E., Smith A. G. and Smith D. G., 1990. *A geologic time scale 1989*. Cambridge University Press, Cambridge, 263 pp.

Harland, W. B., Holland, C. H., House, M. R., Hughes, N. F., Reynolds, A. B., Rudwick, M. J. S., Satterthwaite, G. E., Tarlo, L. B. H. and Willey, E. S. (Eds.) (1967). *The fossil record*. Geological Society, London, pp. 1–828.

Harris, T. M. (1932a). Fossil flora of Scoresby Sound, Part 2. *Medd. Gronland* **85**, 1–112.

Harris, T. M. (1932b). The fossil flora of Scoresby Sound, East Greenland, Part 3, Caytoniales and Bennettitales. *Medd. Gronland* **85**, 5, 5–133, 19 pls.

Harris, T. M. (1933). A new member of the Caytoniales. *New Phytol.* **32**, 97–114.

Harris, T. M. (1940). *Caytonia. Ann. Bot.* **4**, 713–34.

Harris, T. M. (1941). *Caytonanthus*, the microsporophyll of *Caytonia. Ann. Bot.* **5**, 47–58.

Harris, T. M. (1944). A revision of *Williamsoniella. Philos. Trans. R. Soc. Lond. ser. B* **231**, 313–28, 2 pls.

Harris, T. M. (1951). The fructification of *Czekanowskia* and its allies. *Philos. Trans. R. Soc. Lond. ser. B.* **235B**, 483–507.

Harris, T. M. (1957). A Liasso-Rhaetic flora from South Wales. *Proc. R. Soc. Lond.* **147B**, 289–308.

Harris, T. M. (1961). *The Yorkshire Jurassic Flora*, Vol. 1, *Thallophyta-Pteridophyta*. British Museum (Nat. Hist.), London, 212 pp.

Harris, T. M. (1962). The occurrence of the fructification *Carnoconites* in New Zealand. *Trans. R. Soc. NZ, Geol.* **1**, 17–27.

Harris, T. M. (1964). *The Yorkshire Jurassic Flora*, Vol. 2, *Caytoniales, Cycadales and Pteridosperms*. Brit. Museum (Nat. Hist.), London, 191 pp., 7 pls.

Harris, T. M. (1969). *The Yorkshire Jurassic flora*, Vol. 3, *Bennettitales*. British Museum (Nat. Hist.), 186 pp., 7 pls.

Harris, T. M. (1974). *Williamsoniella lignieri*: its pollen and the compression of spherical pollen grains. *Palaeontology* **17**, 125–48, 1 pl.

Harris, T. M. (1976). The Mesozoic gymnosperms. *Rev. Palaeobot. Palynol.* **21**, 119–34.

Harris, T. M. (1979). *The Yorkshire Jurassic flora*, Vol. 5, *Coniferales*. British Museum (Nat. Hist), London, 166 pp., 7 pls.

Harris, T. M., Millington W. and Miller, J. (1974). *The Yorkshire Jurassic flora*, Vol. 4, *Ginkgoales and Czekanowskiales*. British Museum (Nat. Hist.), London, 150 pp., 8 pls.

Hawksworth, D. L. (Ed.) (1991). Improving the stablity of names: needs and options. *Regnum vegetabile* **123**, 358 pp.

Heer, O. (1876). *Flora fossilii Helvetias*. Zurich.

Herendeen, P. S. (1991a). Lauraceous wood from the mid-Cretaceous Potomac Group of Eastern North America: *Paraphyllanthoxylon marylandense* sp. nov. *Rev. Palaeobot. Palynol.* **69**, 277–90.

Herendeen, P. S. (1991b). Charcoalified angiosperm wood from the Cretaceous of Eastern North America and Europe. *Rev. Palaeobot. Palynol.* **70**, 225–39.

Herendeen, P. S. and Dilcher, D. L. (1992). *Advances in legume systematics*, Part 4, *The fossil record*. Royal Botanic Gardens, Kew, 336 pp.

Herngreen, G. R. W. and Chlonova, A. F. (1981). Cretaceous microfloral provinces. *Pollen Spores* 23, 441–555.

Herngreen, G. R. W. and Duenas Jimenez, H. (1990). Dating of the Cretaceous Une Formation, Colombia and the relationship with the Albian–Cenomanian African–South American microfloral province. *Rev. Palaeobot. Palynol.* 66, 345–59.

Heywood, V. H. (Ed.) (1978). *Flowering plants of the world*. Oxford University Press, Oxford, 336 pp.

Hickey, L. J. (1978). The origin of major features of angiosperm leaf architecture in the fossil record. *Courier Forschungsinstitut Senckenberg* 30, 27–34.

Hickey, L. J. (1984). Road logs and stops, Part 1. In Frederiksen, N. O. and Krafft, K. (Eds.), *Cretaceous and Tertiary stratigraphy, paleontology and structure, south-western Maryland and northeastern Virginia, Field trip volume and guide-book*. American Association of Stratigraphic Palynologists, Dallas, pp. 193–209.

Hickey, L. J. and Doyle, J. A. (1977). Early Cretaceous fossil evidence for angiosperm evolution. *Bot. Rev.* 43, 2–104.

Hochuli, P. A. (1981). North Gondwanan floral elements in Lower Cretaceous sediments of the Southern Alps (Southern Switzerland, Northern Italy). *Rev. Palaeobot. Palynol.* 35, 337–58.

Hua, R. (1991). *Early Cretaceous angiospermous pollen from Eren Basin, Nei Mongol*. Geol. Publishing House, Beijing, 66 pp. 12 pl.

Hughes, N. F. (1955). Wealden plant microfossils. *Geol. Mag.* 92, 201–17.

Hughes, N. F. (1961). Further interpretation of *Eucommiidites* Erdtman 1948. *Palae-ontology*, 4, 292–9.

Hughes, N. F. (1973). Mesozoic and Tertiary distributions and problems of land plant evolution. In Hughes, N. F. (ed.) Organisms and continents through time. *Spec. Pap. Palaeontol.* 12, 189–98.

Hughes, N. F. (1976). *Palaeobiology of angiosperm origins*. Cambridge University Press, Cambridge, 242 pp.

Hughes, N. F. (1977). Palaeo-succession of earliest angiosperm evolution. *Bot. Rev.* 43, 105–27.

Hughes, N. F. (1989). *Fossils as information*. Cambridge University Press, Cambridge, 136 pp.

Hughes, N. F. (1992). Mesozoic gymnosperms and period classifications. *Curr. Sci.* Allahabad, 61, 630–3.

Hughes, N. F. and Croxton, C. A. (1973). Palynologic correlation of the Dorset Wealden. *Palaeontology* 16, 567–601.

Hughes, N. F., Drewry, G. E. and Laing, J. F. (1979). Barremian earliest angiosperm pollen. *Palaeontology* 22, 513–35, 8 pls.

Hughes, N. F. and McDougall, A. B. (1987). Records of angiospermid pollen entry into the English Early Cretaceous succession. *Rev. Palaeobot. Palynol.* 50, 255–72.

Hughes, N. F. and McDougall, A. B. (1989). New Wealden correlation for the Wessex Basin. *Proc. Geol. Assoc.* 101, 85–90.

Hughes, N. F., McDougall, A. B. and Chapman, J. L. (1991). Exceptional new record of Cretaceous Hauterivian angiospermid pollen from Southern England. *J. Micropalaeontol.* 10, 75–82.

Hughes, N. F. and Moody-Stuart, J. C. (1969). A method of stratigraphic correlation using Early Cretaceous miospores. *Palaeontology* **12**, 84–111.

Hutchinson, J. (1964). *The genera of flowering plants*, 2 Volumes. Oxford University Press, Oxford.

Jansonius, J. (1992). New requirements in the International Code of Botanical Nomenclature. *Rev. Palaeobot. Palynol.* **71**, 297–8.

Jarzembowski, E. A. (1984). Early Cretaceous insects from Southern England. *Mod. Geol.* **9**, 71–93.

Jarzembowski, E. A. (1991). New insects from the Weald Clay of the Weald. *Proc. Geol. Assoc.* **102**, 93–108.

Jarzen, D. M. (1978). Some Maestrichtian palynomorphs and their phytogeographical and paleoecological implications. *Palynology*, **2**, 29–38.

Jarzen, D. M. and Dettmann, M. E. (1992). Structure and form of Austral Cretaceous Normapolles-like pollen. *Geobios* **25**, 569–83.

Jell, P. A. and Duncan, P. M. (1986). Invertebrates, mainly insects, for the fresh water, Lower Cretaceous, Koonwarra Fossil Beds (Konemburra Group), South Gippsland, Victoria. *Mem. Assoc. Austral. Palaeontol.* **3**, 111–205.

Juhasz, M. and Goczan, F. (1985). Comparative study of Albian monosulcate angiosperm pollen grains. *Acta Biol. Szeged* **31**, 147–72.

Jung, W. W. (1968). *Hirmerella muensteri* (Schenk) Jung nov. comb. eine bedeutsame Konifere des Mesozoikums. *Palaeontographica* **122 B**, 55–93.

Kedves, M. and Pardutz, A. (1992). Transmission electron microscopy of partially dissolved exines of different bisaccate gymnosperm pollen grains. *Plant Cell Biol. Dev. Szeged*, **3**, 38–66.

Kemp, E. M. (1968). Probable angiosperm pollen from the British Barremian to Albian strata. *Palaeontology*, **11**, 421–34.

Kendall, M. W. (1947). On five species of *Brachyphyllum* from the Jurassic of Yorkshire and Wiltshire. *Ann. Mag. Nat. Hist. ser.* 11, **14**, 225–51.

Kendall, M. W. (1949). A Jurassic member of the Araucariaceae. *Ann. Bot.* **13**, 151–61.

Kerth, M. and Hailwood, E. A. (1988). Magnetostratigraphy of the Lower Cretaceous Vectis Formation (Wealden Group) on the Isle of Wight, Southern England. *J. Geol. Soc. Lond.* **145**, 351–60.

Kirchner, M. and Müller, A. (1992). *Umkomasia franconica* n.sp. und *Pteruchus septentrionalis* n. sp., Fruktifikationen von *Thinnfeldia* Ettingshausen. *Palaeontographica* **224 B**, 63–73.

Klaus, W. W. (1960). Sporen der karnischen stufe der Ostalpinen Trias. *Jahrb. Geol. Bundesanst*, Wien, **5**, 107–84.

Knobloch E. and Mai, D. H. (1986). Monographie der Früchte und Samen aus der Kreide von Mitteleuropa. *Rozpravy ustredn. ustavu geol.* **47**, 1–219.

Kohring, R. (1991). Lizard egg shells from the Lower Cretaceous of Cuenca province, Spain. *Palaeontology* **34**, 237–240.

Kotova, I. Z. (1978). Spores and pollen from Cretaceous deposits of the Eastern North Atlantic Ocean, Deep Sea Drilling Project Leg 41, Sites 367 and 370. *Init. Rep. Deep Sea Drill. Proj.* **41**, 841–81.

Kovach, W. L. and Dilcher, D. L. (1988). Megaspores and other dispersed plant remains from the Dakota Formation (Cenomanian) of Kansas, USA. *Palynology* **12**, 89–119.

Krassilov, V. A. (1970). Approach to the classification of Mesozoic 'Ginkgoalean' plants from Siberia. *Palaeobotanist* **18**, 12–19.

Krassilov, V. A. (1972). [Mesozoic flora of Burei (Ginkgoales and Czekanowskiales).] (In Russian.) *Far-East Geol. Inst. Acad. Sci. USSR*, Moscow, 160 pp.

Krassilov, V. A. (1975). Dirhopalostachyaceae, a new family of proangiosperms and its bearing on the problem of angiosperm ancestry. *Palaeontographica* **153 B**, 100–10.

Krassilov, V. A. (1977). Contribution to the knowledge of the Caytoniales. *Rev. Palaeobot. Palynol.* **24**, 155–78.

Krassilov, V. A. (1981). Changes of Mesozoic vegetation and the extinction of dinosaurs. *Palaeogeogr. Palaeoclimatology, Palaeoecol.* **34**, 207–24.

Krassilov, V. A. (1982a). Early Cretaceous flora of Mongolia. *Palaeontographica* **181 B**, 1–43.

Krassilov, V. A. (1982b). On the ovuliferous organ of *Hirmerella*. In *Pant Commemorative Volume, Phyta* pp. 141–4.

Krassilov, V. A. (1984). New palaeobotanical data on origin and early evolution of angiospermy. *Ann. MO Bot. Gard.* **71**, 577–92.

Krassilov, V. A. (1985). *Cretaceous period; evolution of earth's crust and biosphere.*] (In Russian.) Nauka, Moscow, 240 pp.

Krassilov, V. A. (1986). New floral structure from the Lower Cretaceous of Lake Baikal area. *Rev Palaeobot. Palynol.* **47**, 9–16.

Krassilov, V. A. (1987). Palaeobotany of the Mesophyticum: state of the art. *Rev. Palaeobot. Palynol.* **50**, 231–54.

Krassilov, V. A. (1990). Fossil links reconsidered. *Proc. 3rd Int. Org. Palaeobot. Conf.*, Melbourne 1988, pp. 7–10.

Krassilov, V. A. (1991). The origin of angiosperms: new and old problems. *Tree* **6**, 215–20.

Krassilov, V. A. and Ash, S. R (1988). On Dinophyton–Protognetalean Mesozoic plant. *Palaeontographica* **208 B**, 33–8.

Krassilov, V. A. and Bugdaeva, E. V. (1982). Achene-like fossils from the Lower Cretaceous of the Lake Baikal area. *Rev. Palaeobot. Palynol.* **36**, 279–95.

Krassilov, V. A. and Bugdaeva, E. V. (1988a). Protocycadopsid pteridosperms from the Lower Cretaceous of Transbaikalie, and the origin of cycads. *Palaeontographica* **208 B**, 27–32, 4 pls.

Krassilov, V. A. and Bugdaeva, E. V. (1988b). Gnetalean plants in the Jurassic of Ust-Balej, East Siberia. *Rev. Palaeobot. Palynol.* **53**, 359–76.

Krassilov, V. A. and Bugdaeva, E. V. (1988c). *Meeusella* and the origin of stamens. *Lethaia* **21**, 425–31.

Krassilov, V. A. and Pacltova, B. (1989). *Asterocelastrus cretacea*, a mid-Cretaceous angiosperm from Bohemia. *Rev. Palaeobot. Palynol.* **60**, 1–6.

Krassilov, V. A. and Rasnitzyn, A. P. (1982). [The unique findings of pollen grains in the guts of Cretaceous Xyelidae]. (In Russian.) *Palaeontol. J. Moscow* **4**, 83–96.

Krassilov, V. A., Shilin, P. V. and Vakrameev, V. A. (1983). Cretaceous flowers from Kazakhstan. *Rev. Paloeobot. Palynol.* **40**, 91–113.

Krassilov, V. A. and Sukatcheva (1979). [Caddisfly cases of the Karkenia (Ginkgoales) seeds in the Lower Cretaceous of Mongolia]. (In Russian.) In *Far-Eastern Palaeofloristics.* Far-East Science Centre, Vladivostok, pp. 119–21.

Kräusel, R. (1949). Koniferen und andere Gymnospermen aus der Trias von Lunz, Nieder–Österreich. *Palaeontographica* **89 B**, 35–78.

Laing, J. F. (1975). Mid-Cretaceous angiosperm pollen from Southern England and Northern France. *Palaeontology* **18**, 775–808.

Litwin, R. J., Traverse, A. and Ash, S. R. (1991). Preliminary palynological zonation of the Chinle Formation, southwestern USA, and its correlation to the Newark Supergroup (eastern USA). *Rev. Palaeobot. Palynol.* **68**, 269–87.

Long, A. G. (1977). Lower Carboniferous ptesidosperm cupules and the origin of angiosperm origin. *Trans. R. Soc. Edin.* **70**, 13–25.

Lorch, J. (1968). Some Jurassic conifers from Israel. *Bot. J. Linn. Soc.* **61**, 177–88.

Lowe, J. (1961). The phylogeny of monocotyledons. *New Phytol.* **60**, 355–87.

Mabberley, D. J. (1984). The optimistic in pursuit of the unrecognisable: a note on the origin of angiosperms. *Taxon* **33**, 77–9.

Manum, S. B., Bose, M. N. and Sawyer, R. T. (1991). Clitellate cocoons in freshwater deposits since the Triassic. *Zool. Scripta* **20**, 347–66.

Martin, W., Gierl, A. and Saedler, H. (1989). Molecular evidence for pre-Cretaceous angiosperm origins. *Nature* **339**, 46–8.

McClammer, J. U. and Crabtree, D. R. (1989). Post-Barremian (Early Cretaceous) to Paleocene palaeobotanical collections in the Western Interior of North America. *Rev. Palaeobot. Palynol.* **57**, 221–32.

McLaren, D. J. and Goodfellow, W. D. (1990). Geological and biological consequences of giant impacts. *Ann. Rev. Earth Planet. Sci.* **18**, 123–71.

Medus, J. (1977). The ultra structure of some *Circumpolles. Grana* **16**, 23–8.

Meeuse, A. D. J. (1979). Why were early angiosperms so successful? A morphological, ecological and phylogenetic approach. *Proc. Koningklijke. Nederlandse Akademie*, ser. C **82**, 343–69.

Meeuse, A. D. J. (1981). Evolution of the Magnoliophyta: current and dissident viewpoints. *Ann. Rev. Plant Sci.* **2**, 393–442, Kalyani Publishers, New Delhi.

Meeuse, A. D. J. (1990). *Flowers and fossils.* Eburon, Delft, 95 pp.

Melville, R. (1962). A new theory of the angiosperm flower. I. The gynoecium. *Kew. Bull.* **16**, 1–50.

Melville, R. (1969). Leaf variation patterns and the origin of angiosperms. *Nature* **224**, 121–5.

Melville, R. (1983). Glossopteridae, Angiospermidae and the evidence of angiosperm origin. *Bot. J. Linn. Soc.* **86**, 279–323.

Metcalfe, C. R. and Chalk, L. (1979). *Anatomy of the dicotyledons*, 2nd edn, Vol. 1, *Systematic anatomy of the leaf and stem.* Oxford Science Publications, Oxford, 276 pp, 18 pls.

Meyen, S. V. (1987). *Fundamentals of palaeobotany.* Chapman and Hall, London, 432 pp.

Meyen, S. V. (1988). Origin of the angiosperm gynoecium by gamo-heterotropy. *Bot. J. Linn. Soc.* **97**, 171–8.

Meyer-Berthaud, B., Taylor, E. L. and Taylor, T. N. (1992). Reconstructing the Gondwana seed-fern *Dicroidium*; evidence from the Triassic of Antarctica. *Geobios* **25**, 341–4.

Muller, J. (1981). Fossil pollen records of extant angiosperms. *Bot. Rev.* **47**, 1–142.

Muller, J. (1984). Significance of fossil pollen for angiosperm history. *Ann. MO Bot. Gard.* **71**, 419–43.

Muller, J., Di Giacomo, E. and van Erwe, A. W. (1987). A palynological zonation of the Cretaceous, Tertiary and Quaternary of Northern South America. *AASP Contributions Series*, **19**, 76 pp.

Nathorst, A. G. (1909). Pälaobotanische Mitteilungen 8. Über *Williamsonia, Wielandia, Cycadocephalus* und *Weltrichia. Kungl. Svenska Vetansk Akad. Handl.* Stockholm, **45**, 4.

Neale, J. W. (1974). Cretaceous. In Rayner, D. H. and Hemingway, J. E. (Eds.) *The geology and mineral resources of Yorkshire.* Yorkshire Geological Society, pp. 225–43.

Oldham, T. C. B. (1976). The plant debris beds of the English Wealden. *Palaeontology* **19**, 437–502.

Osborn, J. M., Taylor, T. N. and Crane, P. R. (1991). The ultrastructure of *Sahnia* pollen (Pentoxylales). *Am. J. Bot.* **78**, 1560–9.

Osborn, J. M., Taylor, T. N. and de Lima, M. R. (1993). The ultrastructure of fossil ephedroid pollen with gnetalean affinities from the Lower Cretaceous of Brazil. *Rev. Palaeobot. Palynol.* **77**, 171–84.

Pacltova, B. (1971). Palynological study of angiosperms in the Peruc Formation (?Albian–Lower Cenomanian) of Bohemia. *Sb. Geol. Ved Paleontol.*, Praha **13**, 105–41.

Pacltova, B. (1977). Cretaceous angiosperms of Bohemia – central Europe. *Bot. Rev.* **43**, 128–41.

Pant, D. D. (1990). Importance of palaeobotany in the study of botany, zoology, geology and archaeology. *Palaeobotanist* **38**, 1–19.

Pant, D. D. and Kidwai, P. F. (1972). The origin and evolution of flowering plants. *J. Indian Bot. Soc.* **56 A**, 242–74.

Pant, D. D. and Nautiyal, D. D. (1984). On the morphology and structure of *Ottokaria zeilleri* sp. nov. – a female fructification of *Glossopteris. Palaeontographica* **193 B**, 127–52.

Patterson, C. (1981). Significance of fossils in determining evolutionary relationships. *Ann. Rev. Ecol. Syst.* **12**, 195–223.

Paul, C. R. C. (1985). The adequacy of the fossil record reconsidered. *Spec. Papers Palaeontol.* **33**, 7–15.

Pedersen, K. R., Crane, P. R. and Friis, E. M. (1989). The morphology and phylogenetic significance of *Vardekloeftia* Harris (Benettitales). *Rev. Palaeobot. Palynol.* **60**, 7–24.

Pedersen, K. R., Crane, P. R., Drinnan, A. N. and Friis, E. M. (1991). Fruits from the mid-Cretaceous of North America with pollen grains of *Clavatipollenites* type. *Grana* **29**, 577–90.

Pellmeyer, O. and Thieu, L. B. (1986). Insect reproduction and floral fragrances; keys to the evolution of the angiosperms. *Taxon,* **35**, 76–85.

Penny, J. H. J. (1988*a*). Early Cretaceous acolumellate semitectate pollen from Egypt. *Palaeontology* **31**, 373–418.

Penny, J. H. J. (1988*b*). Early Cretaceous striate tricolpate pollen from the Borehole Mersa Matruh 1, North West Desert, Egypt. *J. Micropalaeontol.* **7**, 201–15.

Penny, J. H. J. (1989). New Early Cretaceous forms of the angiosperm pollen genus *Afropollis* from England and Egypt. *Rev. Palaeobot. Palynol.* **58**, 289–99.

Penny, J. H. J. (1991). Early Cretaceous angiosperm pollen from the Borehole Mersa Matruh 1, North-west Desert, Egypt. *Palaeontographica,* **222 B**, 31–88, 14 pls.

Penny, J. H. J. (1992). The relevance of the Early Cretaceous palynology of Egypt to biostratigraphy and reconstruction of angiosperm palaeolatitudinal migrations. *Cret. Res.* **13**, 369–78.

Plumstead, E. P. (1952). Descriptions of two new genera and six new species of fructifications borne on *Glossopteris* leaves. *Trans. Geol. Soc. S. Africa* **55**, 281–328.

Plumstead, E. P. (1956). Bisexual fructifications borne on *Glossopteris* leaves from South America. *Palaeontographica* **100 B**, 1–25.

Plumstead, E. P. (1958). Further fructifications of the Glossopteridae and a provisional classification based on them. *Trans. Geol. Soc. S. Africa* **61**, 57–76.

Pocock, S. A. J. and Vasanthy, G. (1988). *Cornetipollis reticulata*, a new pollen with angiospermid features from Upper Triassic (Carnian) sediments of Arizona (USA), with notes on *Equisetosporites*. *Rev. Palaeobot. Palynol.* **55**, 337–56.

Pocock, S. A. J., Vasanthy, G. and Venkatachala, B. S. (1990). Pollen of Circumpolles – an enigma or morphotrends showing evolutionary adaptation. *Rev. Palaeobot. Palynol.* **65**, 179–93.

Pons, D., Lauverjat, J. and Broutin, J. (1980). Paléoclimatologie comparée de deux gisements du Crétace superieur d'Europe occidentale. *Mém. Soc. Geol. France* **139**, 151–8.

Poort, R. J. and Kerp, J. H. F. (1990). Aspects of Permian palaeobotany and palynology. XI. On the recognition of true peltasperms in the Upper Permian of Western and Central Europe and a reclassification of species formerly included in *Peltaspermum* Harris. *Rev. Palaeobot. Palynol.* **63**, 197–225.

Rasnitsyn, A. P. (1984). The first find of a Jurassic butterfly. *Dokl. Akad. Nauk USSR* (Earth Sci. Section) **269**, 174–7.

Rawson, P. F., Curry, D., Dilley, F. C., Hancock, J. M., Kennedy, W. J., Neale, J. W., Wood, C. J. and Worssam, B. C. (1978). A correlation of Cretaceous rocks in the British Isles. *Geol. Soc. Lond. Spec. Rep.* **9**, 70 pp.

Rees, P. McA. (1993). Caytoniales in Early Jurassic floras from Antarctica. *Geobios* **26**, 33–42.

Regali, M. S. P. (1989). *Tucanopollis*, a new genus of early angiosperms. *Bol. Geosci. Petrobras*, Rio de Janeiro, **3**, 395–402.

Regali, M. S., Uesugui, N. and Santos, A. S. (1974). Palinológia dos sedimentos meso-cenozoicos do Brasil. *Bol. Tec. Petrobras.* **17**, 177–91, 263–301.

Reid, E. M. and Chandler, M. E. J. (1933). *The flora of the London Clay*. British Museum (Nat. Hist.), London, 561 pp.

Retallack, G. (1977). Reconstructing Triassic vegetation of eastern Australasia; a new approach to the biostratigraphy of Gondwanaland. *Alcheringa* **1**, 247–77.

Retallack, G. (1986). The fossil record of soils. In Wright, V. P. (Ed.) *Paleosols: their recognition and interpretation*. Blackwell, Oxford, pp. 1–57.

Retallack, G. and Dilcher, D. L. (1981*a*). Arguments for a glossopterid ancestry of angiosperms. *Palaeobiology* **7**, 54–67.

Retallack, G. and Dilcher, D. L. (1981*b*). Early angiosperm reproduction: *Prisca reynoldsii* gen. et. sp. nov. from mid-Cretaceous coastal deposits in Kansas, USA. *Palaeontographica* **179 B**, 103–37.

Retallack, G. and Dilcher, D. L. (1986). Cretaceous angiosperm invasion of North America. *Cret. Res.* **7**, 227–52.

Reymanovna, M. (1973). The Jurassic flora of Grojec near Krakow in Poland. Part 2. Caytoniales and the anatomy of *Caytonia*. *Acta Palaeobot.* **14**, 46–87.

Reyre, Y. (1970). Stereoscan observations on the pollen genus *Classopollis* Pflug 1953. *Palaeontology* **13**, 303–22.

Rich, T. H. V. and Rich, P. V. (1989). Polar dinosaurs and biotas of the Early Cretaceous of Southeastern Australia. *Nat. Geogr. Res.* **5**, 15–53.

Sahni, B. (1948). The Pentoxyleae: a new group of Jurassic gymnosperms from the Rajmahal Hills of India. *Bot. Gaz.* **110**, 47–80.

Salami, M. B. (1985). Upper Senonian and Lower Tertiary pollen grains from Southern Nigeria sedimentary basin. *Rev. Esp. Micropal.* **17**, 5–26.

Salami, M. B. (1990). Palynomorph taxa from the 'Lower Coal Measures' deposits of Anambra Trough, Southwestern Nigeria. *J. Afr. Earth Sci.* **11**, 135–50.

Samylina, V. A. (1968). Early Cretaceous angiosperms of the Soviet Union based on leaf and fruit remains. *Bot. J. Linn. Soc.* **61**, 207–18.

Samylina, V. A. (1992). The stomatal types of leaves of Mesozoic Ginkgophyta. *Fourth Int. Org. Palaeobot. Conf.* Paris, August 1992, Abstracts, p. 144.

Sanz, J. L. and Buscalioni, A. D. (1992). A new bird from the Early Cretaceous of Las Hoyas, Spain, and the early radiation of birds. *Palaeontology* **35**, 829–45.

Schopf, J. M., Wilson, L. R. and Bentall, R. (1944). An annotated synopsis of paleozoic fossil spores and the definition of generic groups. *Illinois State, Geol. Surv. Rep. Invest.* **91**.

Schrank, E. (1983). Scanning electron and light microscopic investigations of angiosperm pollen from the Lower Cretaceous of Egypt. *Pollen Spores* **25**, 213–42.

Schulz, E. (1967). Sporenpaläontologische Untersuchungen der rätoliassischer Schichten in Zentralteil des Germaninschen Beckens. *Paläont. Abh.*, Ser. 2 **3**, 427–633.

Schweitzer, H. T. (1977). Die Räto-Jurassischen Floren des Iran und Afghanistans. 4. Die Rätische Zwilterblüte *Irania hermaphroditica* nov. spec. und ihre Bedeutung für die phylogenie der Angiospermen. *Palaeontographica* **161 B**, 97–145.

Scott, R. A. (1954). Fossil fruits and seeds from the Eocene Clarno Formation of Oregon. *Palaeontographica* **96 B**, 66–97.

Seward, A. C. (1894). *Catalogue of Mesozoic Plants, The Wealden Flora*, Part I. British Museum (Nat. Hist.), London, 179 pp. 11 pls.

Seward, A. C. (1895). *The Wealden flora*, Part 2 *Gymnospermae*. British Museum (Nat. Hist.), London. 259 pp. 20 pls.

Seward, A. C. 1898–1919. *Fossil plants*, 4 Vols. Cambridge University Press, Cambridge.

Seward, A. C. (1926). The Cretaceous plant-bearing rocks of Western Greenland. *Philos. Trans. R. Soc. Lond.* **215**, 57–175.

Shang, Yu-Ke and Wang, Shu-Ying (1992). Palynomorph assemblages from the Yingchan Formation, Jintai, Jilin. *Acta Micropalaeontol. Sinica* **8**, 91–100.

Shields, O. (1988). Mesozoic history and neontology of Lepidoptera in relation of Trichoptera, Mecoptera, and angiosperms *J. Palaeontol.* **62**, 251–8.

Singh, G. (1988). History of aridland vegetation and climate; a global perspective. *Biol. Rev.* Cambridge, 63, 159–95.

Sloan, L. C. and Barron, E. J. (1990). Equable climates during earth history. *Geology* **18**, 489–92.

Smith, A. G., Hurley, A. M. and Briden, J. C. (1981). *Phanerozoic paleocontinental world maps*. Cambridge University Press, Cambridge, 102 pp.

Solms-Laubach, H., 1891. *Fossil botany*. Engl. translat. by H. E. F. Garnsey, revised by I. B. Balfour. Clarendon Press, Oxford, 401 pp. (German edition 1887.)

Spicer, R. A. (1986). Comparative leaf architectural analysis of Cretaceous radiating angiosperms. In Spicer, R. A. and Thomas, B. A. (Eds.) *Systematic and taxonomic approaches in palaeobotany. Syst. Assoc. Spec.* Vol, **31**, 223–33.

Spicer, R. A. (1989). Physiological characteristics of land plants in relation to environment through time. *Trans. R. Soc. Edin. Earth Sci.* **80**, 321–29.

Spicer, R. (1991). Plant taphonomic processes. In Allison, P. A. and Briggs, D. E. G. (Eds.) *Topics in geobiology*, vol. 9. Plenum Press, New York, pp. 71–113.

Spicer, R. A. and Chapman, J. L. (1990). Climatic change and the evolution of high-latitude terrestrial vegetation and floras. *Trends Ecol. and Evol.* **5**, 279–84.

Spicer, R. A. and Corfield, R. M. (1992). A review of terrestrial and marine climates in the Cretaceous with implications for modelling the 'Greenhouse Earth'. *Geol. Mag.* **129**, 169–80.

Spicer, R. A., Parrish, J. T. and Grant, P. R. (1992). Evolution of vegetation and coal-forming environments in the Late Cretaceous of the North Slope of Alaska. In McCabe, P. J. and Parrish, J. T. (Eds.) *Controls on the distribution and quality of Cretaceous Coals. Geol. Soc. Am. Spec. Paper* **267**, 177–92.

Spicer R. A., Rees, P. McA. and Chapman, J. L. (1993). Cretaceous phytogeography and climatic signals. *Philos. Trans. R. Soc. Lond. ser. B.* **341**, 277–86.

Sporne, K. R. (1974). *The morphology of angiosperms.* Hutchinson, London, 207 pp.

Sporne, K. R. (1980). Re–investigation of character correlations among dicotyledons. *New Phytol.* **85**, 419–49.

Sporne, K. R. (1982). The Advancement Index vindicated. *New Phytol.* **91**, 137–45.

Srivastava, S. K. (1976). The fossil pollen genus *Classopollis. Lethaia* **9**, 437–57.

Srivastava, A. K. (1991). Evolution tendency in the venation pattern of Glossopteridales. *Geobios, Lyon* **24**, 383–6.

Stebbins, G. L. (1965). The probable growth habit of the earliest flowering plants. *Ann. MO Bot. Gard.* **52**, 457–68.

Stebbins, G. L. (1974). *Flowering plants. Evolution above the species level.* Harvard University Press, Cambridge, MA., 399 pp.

Stewart, W. N. (1983). *Paleobotany and the evolution of plants.* Cambridge University Press, Cambridge, 405 pp.

Stewart, W. N. and Rothwell, G. (1993). *Paleobotany and the evolution of plants.* Cambridge University Press, Cambridge, 535 pp.

Stopes, M. C. (1915). *Catalogue of Cretaceous plants in the British Museum (Nat. Hist.)*, Part 2 *Lower Greensand (Aptian) plants of Britain.* British Museum (Nat. Hist.), London, 360 pp. 32 pls.

Takhtajan, A. (1969). *Flowering plants – origin and dispersal.* Transl. by C. Jeffery. Oliver and Boyd, Edinburgh, 310 pp.

Takhtajan, A. (1976). Neoteny and the origin of flowering plants. In Beck C. B. (Ed.) *Origin and early evolution of angiosperms.* Columbia University Press, New York, pp. 207–19.

Taylor, D. W. (1991). Angiosperm ovules and carpels: their characters and polarities, distribution in basal clades, and structural evolution. *Postilla*, New Haven, **208**, 1–40.

Taylor, D. W. and Hickey, L. J. (1990). An Aptian plant with attached leaves and flowers: implications for angiosperm origin. *Science*, **247**, 702–4.

Taylor, D. W. and Hickey, L. J. (1992). Phylogenetic evidence for the herbaceous origin or angiosperms. *Plant Syst. Evol.* **180**, 137–56.

Taylor, E. L. and Taylor, T. N. (1992). Reproductive biology of the Permian Glosspteridales and their suggested relationship to flowering plants. *Proc. Nat. Acad. Sci. USA*, **89**, 11495–7.

Taylor, T. N. (1981). *Paleobotany.* McGraw-Hill, New York, 589 pp.

Taylor, T. N. and Alvin, K. L. (1984). Ultrastructure and development of Mesozoic pollen *Classopollis. Am. J. Bot.* **71**, 575–87.

Taylor, T. N. and Archangelsky, S. (1985). The Cretaceous pteridosperms *Ruflorinia* and *Ktalenia* and implications on cupule and carpel evolution. *Am. J. Bot.* **72**, 1842–53.

Taylor, T. N. and Taylor, E. L. (1993). *The biology and evolution of fossil plants.* Prentice-Hall, 982 pp.

Taylor, T. N., Zavada, M. S. and Archangelsky, S. (1987). The ultrastructure of *Cyclusphaera psilata* from the Cretaceous of Argentina. *Grana* **26**, 74–80.

Thanikaimoni, G. (1986). Pollen apertures: form and function. In Blackmore, S. and Ferguson, I. K. (Eds.) *Pollen and spores: form and function.* Academic Press, London, pp. 119–136.

Thomas, B. A. and Spicer, R. A. (1986). *The evolution and palaeobiology of land plants.* Croom Helm, London, 309 pp.

Thomas, H. H. (1915). On *Williamsoniella* a new type of Bennettitalean flower. *Phil. Trans. R. Soc. Lond. ser. B* **207**, 113–48, 3 pls.

Thomas, H. H. (1925). The Caytoniales, a new group of angiospermous plants from the Jurassic rocks of Yorkshire. *Phil. Trans. R. Soc. Lond. ser. B* **213**, 299–363.

Thomas, H. H. (1933). On some pteridospermous plants from the Mesozoic rocks of South Africa. *Philos. Trans. R. Soc. Lond. ser. B* **222**, 193–265.

Thomas, H. H. (1934). The nature and origin of the stigma. *New Phytol.* **33**, 173–98.

Thomas, H. H. (1958). Lidgettonia, a new type of fertile *Glossopteris. Bull. Brit. Mus. (Nat. Hist.) Geol.* **3**, 179–89.

Thomas, H. H. and Bancroft, N. (1913). On the cuticles of some recent and fossil Cycadean fronds. *Trans. Linn. Soc.* **8**, 155–204.

Thomas, H. H. and Harris, T. M. (1960). Cycadean cones of the Yorkshire Jurassic. *Senckenberg. leth.* **41**, 139–61.

Thorne, R. F. (1976). Phylogenetic classification of the angiosperms. *Evol. Biol.* **9**, 35–106.

Tidwell, W. D. and Medlyn, D. A. (1992). Short shoots from the Upper Jurassic Morrison Formation of Utah, Wyoming and Colorado, USA. *Rev. Palaeobot. Palynol.* **71**, 219–38.

Tidwell, W. D., Simper, A. D. and Thayn, G. F. (1977). Additional information regarding the controversial Triassic plant, *Sanmiguelia. Palaeontographica* **163 B**, 143–51.

Tiffney, B. H. (1984). Seed size, dispersal syndromes and the rise of the angiosperms: evidence and hypothesis. *Ann. MO Bot. Gard.* **71**, 551–76.

Tomlinson, P. B. (1961). Palmae. In Metcalf, C. R. (Ed.) *Anatomy of the monocotyledons*, Vol. II. Clarendon Press, Oxford, 453 pp.

Townrow, J. A. (1960). The Peltaspermaceae, a Pteridosperm family of Permian and Triassic age. *Palaeontology* **3**, 333–61.

Townrow, J. A. (1962a). On *Pteruchus*, a microsporophyll of the Corystospermaceae. *Bull. Brit. Mus. (Nat. Hist.)* **6**, 2, 287–320.

Townrow, J. A. (1962b). On some bisaccate pollen grains of Permian to Middle Jurassic age. *Grana Palynol.* **3**(2), 13–44.

Tralau, H. (1967). The phytogeographic evolution of the Genus *Ginkgo* L. *Botaniska. Notiser.*, Lund, **120**, 409–22.

Traverse, A. (1988). *Paleopalynology*. Unwin Hyman, Boston, 600 pp.

Trevisan, L. (1971). *Dicheiropollis*, a pollen type from Lower Cretaceous sediments of Southern Tuscany (Italy). *Pollen Spores* **13**, 591–6.

Trevisan, L. (1980). Ultrastructural notes and considerations on *Ephedripites, Eucommiidites* and *Monosulcites* pollen grains from Lower Cretaceous sediments of Southern Tuscany, Italy. *Pollen Spores* **22**, 85–132.

Trevisan, L. (1988). Angiosperm pollen (monosulcate-trichotomosulcate phase) from very early Lower Cretaceous rocks of Southern Tuscany, Italy. *Seventh Int. Palynol. Congr., Brisbane, Abstracts*, p. 165.

Trevisan, L. (1994). Angiospermid pollen (monosulcate–trichotomosulcate phase from Early Cretaceous of Southern Tuscany, Italy. Unpublished report.

Trincao, P. and Pais, J. (1992). Stratigraphic distribution of angiosperm pollen in the Lower Cretaceous of Portugal. *Fourth Conf. Int. Org. Paleobot. Paris, Abstracts* p. 168.

Tripathi, A. and Tiwari, R. S. (1991). Early Cretaceous angiosperm pollen from the intertrappean beds of Rajmahal Basin, Bihar. *Palaeobotanist* **39**, 50–6.

Upchurch, G. R. Jr (1984). Cuticle evolution in Early Cretaceous angiosperms from the Potomac Group of Virginia and Maryland. *Ann. MO Bot. Gard.* **71**, 522–50.

Upchurch, G. R. and Wolfe, J. A. (1987). Mid-Cretaceous to Early Tertiary vegetation and climate: evidence from fossil leaves and woods. In Friis, E. M., Chaloner, W. G. and Crane, P. R. (Eds.) *The origin of angiosperms and their biological consequences*. Cambridge University Press, Cambridge, pp. 75–105.

Vakhrameev, V. A. (1966). Jurassic floras of the USSR. *Palaeobotanist*, **14**, 118–23.

Vakhrameev, V. A. (1988). [Jurassic and Cretaceous floras and climates of the earth.] (In Russian.) *Trudy Geol. Inst. Acad. Sci. USSR*, **430**, 1–210.

Vakhrameev, V. A. (1991). *Jurassic and Cretaceous floras and climates of the earth.* English edn, Hughes, N. F. (Ed.). Cambridge University Press, Cambridge, 318 pp.

Van Konijnenburg-van Cittert, J. H. A. (1971). *In situ* gynmosperm pollen from the Middle Jurassic of Yorkshire. *Acta. Bot. Neerl.* **20**, 1–96, 16 pls.

Van Konijnenburg-van Cittert, J. H. A. (1987). New data on *Pagiophyllum maculosum* Kendall and its male cone from the Jurassic of North Yorkshire. *Rev. Palaeobot. Palynol.* **51**, 95–105.

Van Konijnenburg-van Cittert, J. H. A. (1992). An enigmatic Liassic microsporophyll, yielding *Ephedripites* pollen. *Rev. Palaeobot. Palynol.* **71**, 239–54.

Velenovsky, I. and Viniklar, L. (1926–1931). Flora cretacea Bohemiae. *Rozpr. Statn. Geol. Ustav. Cesk. Rep.*, pp. 1–345.

Vishnu–Mittre (1953). A male flower of the Pentoxyleae with remarks on the structure of the female cones of the group. *Palaeobotanist* **2**, 75–84.

Walker, J. W. (1974). Aperture evolution in the pollen of primitive angiosperms. *Am. J. Bot.* **61**, 1112–37.

Walker, J. W., Brenner, G. J. and Walker, A. G. (1983). Winteraceous pollen in the Lower Cretaceous of Israel: early evidence of a Magnolialean angiosperm family. *Science* **220**, 1273–5.

Walker, J. W. and Walker, A. G. (1984). Ultrastructure of Lower Cretaceous angiosperm pollen and the origin and early evolution of flowering plants. *Ann. MO Bot. Gard.* **71**, 464–521.

Walker, J. W. and Walker, A. G. (1986). Ultrastructure of Lower Cretaceous angiosperm pollen and its evolutionary implications. In Blackmore, S. and Ferguson, I. K. (Eds.) *Pollen and spores; form and function*. Academic Press, London, pp. 203–17.

Ward, L. F., Fontaine, W. M., Bibbins, A. and Wieland, G. R. (1905). Status of Mesozoic floras of the United States. *U.S. Geol. Surv. Monogr.* **48**, 1–616.

Watson, J. (1977). Some Lower Cretaceous conifers of the Cheirolepidiaceae from the USA and England. *Palaeontology* **20**, 715–49.

Watson, J. (1983). A new species of the conifer *Frenelopsis* from the Cretaceous of Sudan. *Bot. J. Linn. Soc.* **86**, 161–7.

Watson, J. (1988). The Cheirolepidiaceae. In Beck, C. B. (Ed.) *Origin and evolution of gymnosperms*. Columbia University Press, New York, pp. 382–447.

Watson, J., Fisher, H. L. and Hall, N. A. (1988). The holotype of the Wealden conifer *Brachyphyllum punctatum* Michael. *Palaeontology* **31**, 1029–31.

Watson, J. and Sincock, C. A. (1992). Bennettitales of the English Wealden. *Monogr. Palaeontograph. Soc. Publ.* 588, 1–228, 23 pls.

Wesley, A. (1966). The fossil flora of the Grey Limestones of Veneto, Northern Italy, and its relationship to the other European floras of simlar age. *Palaeobotanist* **14**, 124–30.

Wettstein, R. (1901). *Handbuch der systematischen Botanik*. Leipzig.

Wettstein, R. (1907). *Handbuch der systematischen Botanik*, 2nd edn. Deuticke, Leipzig und Wien.

Whalley, P. E. S. (1985). The systematics and palaeogeography of the Lower Jurassic insects of Dorset, England. *Bull. Br. Mus. (Nat. Hist.) Geol*, **39**, 107–89.

White, H. J. O. (1921). A short account of the geology of the Isle of Wight. *Mem. Geol. Surv. Gt Br.*, 1–219.

White, M. E. (1986). *The greening of Gondwana*. Reed Books, Sydney.

Wieland, G. R. (1906). *American fossil cycads*, Vol. 1. Carnegie Institution, Washington, pp. 1–295.

Wieland, G. R. (1916). *American fossil cycads*, Vol. 2. Carnegie Institution, Washington, pp. 1–277.

Wieland, G. R. (1934). Fossil cycads, with special reference to *Raumeria reichenbachiana* Goeppert, of the Zwinger of Dresden. *Palaeontographica* 79 **B**, 85–130.

Wing, S. L. and Tiffney, B. H. (1987). Interactions of angiosperms and herbivorous tetrapods through time. In Friis, E. M., Chaloner, W. G. and Crane, P. R. (Eds.) *The origins of angiosperms and their biological consequences*. Cambridge University Press, Cambridge, pp. 203–24.

Wolfe, J. A. (1978). A paleobotanical interpretation of Tertiary Climates in the Northern Hemisphere. *Am. Sci.* **66**, 694–703.

Wolfe, J. A. (1985). Distribution of major vegetational types during the Tertiary. In Sundquist, E. D. and Broecker, W. S. (Eds.) *The carbon cycle and atmospheric CO_2: natural variations Archaean to present. Am. Geophys. Union Geophys. Monogr.* **32**, 357–75.

Wolfe, J. A. (1987). Late Cretaceous–Cenozoic history of deciduousness and the Terminal Cretaceous event. *Paleobiology* **13**, 217–26.

Wolfe, J. A. and Tanai, T. (1987). Systematics, phylogeny and distribution of Acer (Maples) in the Cenozoic of Western North America. *J. Fac. Sci. Hokkaido Univ.* ser. 4 Geol. **22**, 1–246.

Wolfe, J. A. and Upchurch, G. R. Jr 1987. North American non-marine climates and vegetation during the Late Cretaceous. *Palaeogeogr. Palaeoclimatol. Palaeoecol.* **61**, 33–77.

Wolfe, J. A. and Wehr, W. (1988). Rosaceous *Chamaebatiaria*-like foliage from the Paleogene of Western North America. *Aliso* **12**, 177–200.

Worssam, B. C. and Ivimey-Cook, H. C. (1971). The stratigraphy of the Geological Survey Borehole at Warlingham, Surrey. *Bull. Geol. Surv. G.B.* **36**, 1–178.

Zavada, M. S. (1990). The ultrastructure of three monosulcate pollen grains from the Triassic Chinle Formation, Western United States. *Palynology* **14**, 41–51.

Zeiller, R. (1900). *Elements de paléobotanique*. Paris, 417 pp.

Zhou, Z. (1991). Phylogeny and evolutionary trends of Mesozoic ginkgoaleans – a preliminary assessment. *Rev. Palaeobot. Palynol.* **68**, 203–16.

Index

Printed in t
By Bookma